MW00907424

LADY WITH THE IRON RING

AN ENGINEER'S MEMOIR OF HOPE, LUCK AND SUCCESS

NATTALIA LEA

Oct. 12, 2019

Dear Sheila ☺
Believe that all things are possible.
Best,
Nattalia Lea

 FriesenPress

Suite 300 - 990 Fort St
Victoria, BC, V8V 3K2
Canada

www.friesenpress.com

Copyright © 2019 by Lea, Nattalia, 1953 –
Lady with the Iron Ring/Nattalia Lea
First Edition — 2019

All rights reserved. No part of this publication may be reproduced, stored
in a retrieval system, or transmitted in any form by any process – electronic,
mechanical, photocopying, recording, or otherwise – without the prior written
permission of the author and Friesen Press. The scanning, uploading, and
distribution of this book via the Internet or via any other means without the
permission of the publisher is illegal and punishable by law. Please purchase
only authorized electronic editions, and do not participate in or encourage
electronic piracy of copyrighted materials. Your support of the author's rights is
appreciated.

Library and Archives Canada Cataloguing in Publication

Cover photo of author by Jana Miko.
Photos and images to accompany this book are posted on
www.fb.me/LadywiththeIronRing

Disclaimer:
The author has lived most of her life before #MeToo. Some readers may find
the subject matter and opinions expressed offensive and politically incorrect.
Reader discretion is advised.

This is a work of creative nonfiction. The events are portrayed to the best of the author's memory. While all the stories in this book are true, some names and identifying details have been changed to protect the privacy of the people involved.

You'll really like this book. Please buy it and share it with your family and friends.

Attention:
Quantity sales are available to companies, educational institutions or writers' groups for reselling, educational purposes, fundraising campaigns, gifts, and subscription incentives. Please contact:

Nattalia Lea – E-mail - comehomebehappy@gmail.com
Facebook – www.facebook.com/nattalia.lea.3
LinkedIn - https://ca.linkedin.com/in/nattalia-lea-3344362b
Twitter - @Ladyw_IronRing
Instagram – nattalia_lea
www.fb.me/LadywiththeIronRing

ISBN
978-1-5255-4600-6 (Hardcover)
978-1-5255-4601-3 (Paperback)
978-1-5255-4602-0 (eBook)

1. BIOGRAPHY & AUTOBIOGRAPHY, WOMEN

2. HISTORY, CANADA, POST-CONFEDERATION (1867 TO PRESENT)

3. TECHNOLOGY & ENGINEERING, ENGINEERING (GENERAL)

Distributed to the trade by The Ingram Book Company

Dedication

This book is dedicated to that part of you that wants to be true to yourself – no matter what others may think of you, especially your parents.

Introduction

If there's one thing certain about our future, you can expect uncertainty. This is happening as you breathe. Look at how technology, politics, and social media are changing our world – the way we think and react to *circumstances*. Who would have thought that in 2019, up to 80 per cent of Canadian engineering work is now being outsourced to developing countries?

In 1978, I was the first woman to graduate in bio-resources engineering from the University of British Columbia. Four other women and over 200 men graduated in engineering alongside me. Who would have thought, that me – the shortest, the smallest and most visible of my female peers, would find success as an engineer in the once all white male oil industry? I graduated with no mentors and no role models to guide me.

I have survived 12 involuntary job terminations in engineering with a freelance writing career in between. More recently, I have worked as a film maker and screenwriter. Adversity has taught me valuable life lessons in serendipity, adaptation, and courage. I wished there was a book like the one I wrote today to read, instead of Richard Bolle's *What Colour is Your Parachute?*

I wrote this book to share with you, the readers, my story - to inspire you and give you the courage to find your passion in life. Then go for it. It is not by chance that you're reading my book. It's not by chance that I've written this book. I really hope you can learn from my mistakes and save yourself a lot of grief. Remember, please try not to take yourself so seriously. LOL

PART ONE –
MY BRAINWASHING YEARS

Chapter 1 –
In My Own Words

A sucker is born every minute. Why would I be an exception? The only difference between others and me, is I am a very considerate being, not like any of those narcissistic sociopaths that I would eventually meet later in life. My mother vaguely recalled that she went to the hospital in the morning. I was born at 4:51 p.m. in Vancouver, Western Canada's largest city – early enough, so my mother had time for a decent supper. I imagine the doctor grabbed me by my legs, turned me upside down and slapped me on my butt. Then I would yell to announce my arrival and show everyone who was in charge now. That's how every black and white television show always portrayed childbirth, during the 1950s.

I was the runt of the pack of four babies my mother had, weighing in around five and one-half pounds. Underweight, the medical authorities may have classified me. I beg to differ. Compact. Like a car, with better fuel efficiency and performance over other models. Middle child. Number three out of four. Psychologist Dr. Leman, who wrote *The Birth Order Book* in 1982, notes that my parents should not expect much from me, as middle children are not typically good at taking orders. This is the good news. Nobody's going expect me to save the planet, find a cure for cancer, or become prime minister.

Most of my early childhood is a blur. Dr. Leman was right about the middle child being ignored in favour of my older and younger siblings. In my defense, I spoke up, probably not in the most rationale sense, as my ability to pontificate or make sense of my universe was rather crude and unrefined. I cried. They ignored me. I cried more. The cycle repeated. They called me a crybaby. I got that. I learned to shut up.

Chapter 2 –
Close Encounters with Men in Black

Parents really need to re-think about having four children, because that means there's two middle children. Like my older brother and me. I am positive it was very hard for my brother to grow up with three sisters. I'm pretty sure he was as dissatisfied with his early childhood, as much as I was. He was also brilliant, probably even more than I. You see, my brother masterminded a heist. I agreed to participate. Probably, because I really didn't have anything better to do that weekday afternoon.

The climb up those stone steps leading to the bungalow of a popular Vancouverite's house along Peugeot Drive, took forever. My agility was no match for my older brother, who dashed ahead of me. I kept telling myself, "Right foot, left foot."

I was scared about slipping on the moss-covered steps. My brother glanced back for a second, and motioned with his left arm to keep moving. I began to sweat in the wine-coloured corduroy pants with matching sweater that I wore every day, no matter what time of the year it was. It was a warm and muggy summer day, but I didn't have summer clothes to wear. The eczema behind the backs of my knees started to itch. Snot was dripping down my nose.

The dark blue convertible with white-walled tires and shiny bumpers that Mr. Hope drove was not parked in the front. We knew no one was home. My brother had studied that block for weeks and targeted this house. All the houses along Peugeot Drive were for families better than ours. The view overlooking the city elevated your soul to dizzying heights. Such a life wasn't in our cards. We were poorer. There wasn't enough food on the dinner table. My two older siblings didn't drink milk because our parents couldn't afford milk. Dinner was always the same – white rice, stir-fried beef, and green beans or broccoli … seven days a week. If I went for seconds, my big brother would pinch my thighs. If I got too close to him, he would pinch me in my tits.

I sat on the damp grass on the hill facing Peugeot Drive. I wiped the snot from my runny nose on the cuff of my sweater and folded it over. My brother sauntered around to the back deck that faced an enchanted garden. His skinny bod slipped through the half opened back window into the kitchen. We expected that window to be open. People didn't have air-conditioning back then. On summer days, they kept windows open. I looked nervously around to see if anyone was coming. If so, we'd scram. My heart raced, as I pulled the strands of grass through my hands. This seemed to take forever. I never thought about what we would do, if the cops came. That never crossed my mind.

In a matter of minutes, my brother ripped the coupons off the fridge. What were we thinking? In broad daylight. We raced down the steps, cashed in our loot at the local convenience store … a red sugar-coated marshmallow strawberry that stuck to the innards of my mouth and a Tootsie Roll wrapped in brown and white waxed paper. My brother guzzled down a bottle of Coca Cola. All that we worked for was gone in a flash.

We went home and pretended that we had just gone to Trafalgar Park to play for the afternoon. We could do that. Under-aged kids in the 1950s left home alone, now considered child neglect by today's standards, was par for the course. Besides, mothers didn't work. Our mother, was the only mom on the block who did. We were also the only non-white family in the neighbourhood for blocks to come.

A week later, a police cruiser pulled up to our house – a black and white one with a "cherry" on top, some big American car model I recall, likely a Ford and if not, a Chrysler. There was a loud knock on the front door, followed by the ring of the doorbell. Two uniformed men in black, constables I believe they were called, stood tall. Not good cop, bad cop. Just two good cops trying to sustain the neighbourhood watch. My mother answered the door. My father noted the cop car and stood beside her. Our parents called us to the living room.

The four of us – younger sister, me, older brother and bigger sister – sat solemnly on the couch. I was nervous, so nervous. I was itching to scratch my eczema, but I couldn't. I sat there frozen, holding my breath. My brother was distressed, but showed no signs of guilt, shame or remorse. He was at ease. I fought for my life not to squirm. My eyes shifted around the room, avoiding the eyes of my father and the cops. One of the two men in black, the older constable uttered, "There's been a break-in at the Hopes' house. We want to know if you know anything about it. They think it was done by some kids."

Our father politely looked at us. He was convinced that none of us were involved. He was believable to the cops. My father had probably been questioned like this before. You could tell. He ran away from home when he was 15, I was told years later.

My eyes failed to look at my father, as we all shook our heads in unison. But I knew from the bottom of my gut, that the cops knew that my brother and I were involved. I knew that somebody saw us. The cops looked at each other with that I'm not sure what's going on look. The older one decided not to press the issue. They stood up, shook hands with my father, and tipped their hats, before heading out to the front door. My mother, the darling lady she was, looked up to these two handsome men in awe. She smiled demurely to them with her back to our father. The older one stopped to talk to our father, "Well, if you do hear of anything, please give me a call."

The cop handed my father his business card. We were dismissed.

My brother and I never spoke about our illegal adventure. I knew what we did was wrong. I lay awake at night wondering when they were going to come and take me away. I wondered what hell we were in, if our father found out. He'd beat us. For sure, that would be a minimum. He'd take that wooden mixing spoon and whip my backside like no tomorrow. Or smack me across my face, so my cheeks hurt, but my voice was silenced. He had so much rage inside him, so much resentment and so much hate. I feared him, more than the men in black. I was five when this happened. Being a pussy wasn't going to serve me well.

Chapter 3 –
White People Don't Like Us

My grandfather Lee Kew, a labourer on my father's side arrived in Victoria, B.C. in 1897. He came from a village in Canton (now Guangzhou). My grandmother Lai Fong on my mother's side was shipped to Vancouver in 1927 from Sun Wai Lau – at least that's what the manifests show.

My father grew up in Chinatown Victoria on Yates Street in a dwelling that has been long buried and in which now resides a Fountain Tire franchise. My mother grew up in Vancouver Chinatown – a real kickass ghetto, with feuding Chinese and Japanese youth knife-bearing gangs. The house she grew up in on Keefer Street in the heart of Chinatown, has also been buried and replaced by a contemporary condo building for seniors' housing. So is the Chinese Presbyterian Church next door, that I vaguely recall seeing as a little girl.

In 1930, there were approximately three thousand persons living in Vancouver Chinatown. It was 90 per cent male; ten men for every woman. Most of the men had wives and family back in China, but many never married or had any children. The prettiest women worked as waitresses in restaurants, openly serving food and sexual favours. You could say that was business as usual.

Chinatown made me queasy. The rundown crowded dwellings always stunk of something – burning incense, dried orange peels, shrimp and weird body parts for consumption. Large crocks were filled with brine and thousand-year-old duck eggs – a taste my Aunties found delightful, but I found revolting. Chinese restaurants like the Ho Ho served bowls brimming with *congee* and noodles. With Vancouver's relentless rain and moisture, the windows at the Ho Ho were constantly fogged up during the winter. Elderly Chinese with puffy squinty eyes and missing teeth shuffled the streets, with their hands clasped behind their hunched over backs. Treated as second-class citizens, the Chinese who came over at the turn of the last century had their hands tied when it came to working for others, resorting to menial jobs that white people didn't want – on the farms and fish canneries, cooking and cleaning.

Living up to the Asian stereotypes of that era, my parents ran a dry cleaning business. When my father got a job working for the government years later, the dry cleaning business was sold to another Chinese immigrant family.

Chinese can be racists, too. White people referred to us as the "yellow peril." My father called the white people - "*lo fun*", literally meaning "white rice". My parents squawked profusely about how lazy white people were, how fat and ugly they were, or how stupid they were at squandering away their money on necessities we couldn't afford. My parents smiled like a pair of Cheshire cats in front of white people, then mocked them when their backs were turned. I didn't understand a word of what they were saying in Chinese, but I could tell from their body language and gestures – whatever they were saying, wasn't very nice.

Our parents isolated us from white people. They had their reasons. It was only seven years before I was born that they were given the right to vote. Ironically, they didn't send us to Chinese school, because they hated that part of their life growing up and the logistics. We lived in a white middle-class neighbourhood off 21st Avenue and Macdonald in the west part of town – where people of Irish, Scottish, British, and German descent resided. That seemed like an ideal location to get away from Chinatown and close to the university.

On the streets, white people were obviously uncomfortable with our presence. They assumed we couldn't speak English. What they didn't know is that I couldn't speak Chinese, not even *Chinglish*, which frustrates me to this very day that perma-tanned comics like Russell Peters or Anjelah Johnson, speak *Chinglish* fluently. So white people often spoke super slowly or yelled at us, like we were deaf and dumb. I fulfilled their self-fulling prophesy by looking dumb and acting deaf.

While little white kids my age got to go to the circus with their grandparents, I was already in one. Stereotypical white mothers of northern European descent, tall with tight bods and naturally occurring platinum blonde hair, lily white skin, the bluest of blue eyes and rose pink cheeks pranced down the sidewalk with their mini-me daughters. Then they eyed me, "Look, there's a ..."

Then their jaws would drop, and hiss at me, "Chi-ne-se girl."

My heart would miss a beat. I'd assume that I had done something wrong and would be burnt at the stake. They looked like they saw an alien not realizing I wasn't one, because I didn't have a probe to shove up their ass. I'd shiver inside, wishing I was a ghost. Their awkwardness would pass. The shock on their face was priceless. I probably should have called 911, but cell phones and 911 didn't exist back then. Meanwhile, I think my epidermal cells on my face just doubled in thickness.

There was no shame for white people to make fun of us. Even the crippled white boy in the corner house at the end of the block had an opinion, "So what's with Chinese people eating with chopsticks? I don't get it."

I gave him the evil eye, but that failed to register in his mushy mind. What I didn't have the guts to tell him is that we ate with forks at home. We only used chopsticks in Chinese restaurants. Eating with those plastic slippery ones, was a skill worth acquiring unless you're on a diet.

I survived kindergarten without any racial repercussions. I am also pleased to report that I also peed in my pants without any negative repercussions, too. That was a deliberate ploy on my part. I was the last little person in class to accidentally pee. I guess everybody else in my class felt equally awkward coming to school, too. I would bond with my peers. I would prove to them that I was one of them. On the morning, I peed in my pants; I victoriously, walked around the classroom with folded paper towels shoved into my shoes. I never thought of myself as being extra special back then, but I could have. My siblings and I would be the only Chinese kids at Trafalgar primary and elementary school.

But things were different in Grade 1, when bullies came out of the closet to correlate well with Darwin's theory on the survival of the fittest. It was morning recess. Proverbial Vancouver rain by the buckets smacked the tin roof that covered the concrete playground. The boys are throwing a tennis ball against a wall. We girls are playing hopscotch. The ball rolls our way. I grab the ball, so it doesn't roll into the water rivulet that zigzags across the concrete. Gunther, the alpha male of the group, also an Aryan nation poster child, approaches. The other boys follow, circling me like a pack of coyotes and chant, "Chinky, Chinky, Chinky China man went downtown, then turned around the corner and his pants fell down."

Laughter ensued. I feel dread in the bowels of my mind. I gather the gumption to whack the ball back at Gunther. Tragically, I miss. Our eyes lock. His blue eyes narrow. He sucks in his mouth, primes himself to spit. Josephine, the tallest girl in Grade 1 and possibly Grade 2, steps in front of me, shielding me from Gunther's glare. Now all the kids are staring.

Josephine calmly tells Gunther to back off. The recess bell rings. Gunther cocks his head up, like he's done absolutely nothing wrong. Josephine consoles me, as I swat the tears streaming down my face. All I hear is the sound of that relentless rain turning that former water rivulet into a major river crossing my feet.

No wonder I went home after school every day and cried myself to sleep.

Chapter 4 –
White Girl Envy

I saw pretty young girls in frilly frocks, curls and waves in their hair with sparkly headbands and barrettes, white ankle socks and black patent leather shoes coming to school. They had blue, green, hazel and grey round eyes with defined eyelids and curvy long eyelashes. Their fine hair was blonde, mousy blonde, auburn, or fifty shades of brown. When the sun shone on their heads, their hair shimmered in unison like a symphony of silk threads. They were taller than me. They had beautiful white skin – some with freckles, some rosy and soft or peaches and cream. They all looked like little dolls that smiled and giggled freely.

I'd come home and look in the mirror. My hair was thick and straight. I had bangs and short hair. I could twirl loose strands around my fingers and hold them tight for a minute and the strands would straighten themselves out at the speed of light. My eyes were brown and almond-shaped, with puffy bags underneath, from crying. My nose was flat. I wanted a white girl's nose – sharper and more pronounced. My face was plain and flat like a pancake. I felt ugly.

I looked down at the clothes I wore to school – hand-me-downs or something that my grandmother picked up at the discount Army & Navy – boxfuls she'd buy in bulk because these were the rejects that nobody wanted to wear. Didn't matter if the shoes were too big, just wear them; even if I got blisters or couldn't run in them. If the skirt was too big, just pinch it in with a safety pin.

I wanted to wear sparkling birthstone rings with adjustable bands and fake coloured stones. The only jewelry I owned was a silver locket, tucked away in my mother's dresser in a coveted blue box stamped 'Birks'. I'm speculating about that box because we didn't have the money to shop there. That locket was only worn about once every leap year. I'd sneak into my parents' room and open the top right dresser drawer, where my mother had all her jewelry organized in little open boxes. I especially liked her fake emerald green bracelet – of a gazillion carats worth of viridian green glass stones set side by side and draped around my wrist like a slinky.

None of us kids ever asked for anything more than we were given because we knew we were poor. The white kids got allowances. We didn't. Entitlement was not part of our vocabulary. There was no point in challenging our parents. If I needed cash, like a nickel to buy a donut for the Red Cross charity, collecting and cashing in empty Coke and Crush glass pop bottles plucked from grimy ditches was more preferable than the previous breaking and entering.

I wanted to eat white people's food – *Dad's* cookies, boxed raisins, cheddar cheese for recess; sandwiches with lettuce and tomato, bacon, pancakes, Kraft blue boxed macaroni and cheese, anything called a casserole; roast beef and gravy, mashed potatoes, apple pie and ice cream.

Once, my white people's food fantasy came true. I crashed a birthday party with my brother. It was some kid in his Grade 2 class. Back in those days, you had to invite everybody in your class to your birthday, even if you didn't know them. My brother got the invitation. My mother didn't even bother to ask the birthday boy's mother if it was kosher to send me. She had to work. Off I went. We had hot dogs, potato chips, cake, Jell-O and ice cream. This was diabetic's hell with all the refined processed food, white flour, and sugar. The hot dog filled me up, leaving little room for the cake, Jell-O and ice cream. The cake was a light fluffy layered cake smothered in vanilla cream icing, piped red roses and rosettes and green leaves. Lucky waxed paper wrapped pennies were hidden in between the layers. I had a hard time believing that the silver pearls were actually edible. I picked the pearls out and moved them to the side of the plate. I couldn't eat the ice cream as fast as it was melting. There I was pigging out, but unlike a cow, I only had one stomach. I looked ruefully at the pile of mush the Jell-O and ice cream had become. My brother winced at me, shook his head and played with the mush. The white kids were chasing each other around the house. They squealed like pigs wrestling in pig poop. They would have been surely been smacked in our house. My brother and I sat quietly in our chairs, gawking at the world going by, while I wondered where the bathroom was. I was sick, my tummy really hurt. I was worried that I was going to fart. Someone would take notice. I wished I could vaporize into thin air. Maybe the food we ate at home wasn't so bad after all. At least it was easier on my system.

I wanted to go to camp for summer vacation and Disneyland to see Mickey Mouse. I wanted a girly girl's room in pink, fluffy comforters, stuffed toys and Barbie dolls. Instead, I made dolls out of brown butcher tape I tore off meagre strips from the kitchen drawer and pilfered Kleenex. If we were lucky, my mother took us to visit her sisters, and I got to build houses out of Lego with my cousins. I wanted to go to church so I could wear a silver cross around my neck and tell everyone that I was a Christian, for real.

I wanted to take piano lessons, prance around in a pink tutu and spin pirouettes on ice. I wanted to be a princess, too. Maybe, I'd meet someone who will love and cherish me, and we'd live happily ever after. I didn't care if I was being realistic or not. Of course, I knew as if this was ever going to be possible. White people didn't like us.

Chapter 5 –
Must Not End Up Like My Mother

People are just being polite when they tell you that you look like your mother. They always say this to you, even if you look more like your father, minus his facial hair and bald spot. They just want you to know that you were cared for enough that your parents kept you and didn't put you up for adoption. Well, that's the feeling I got. Oh, how endearing that I'm a clone of my mother. She would have been insulted the number of times I was told that I looked like my mother, as she was like no other I knew of. She was extra SPECIAL.

My mother wanted to be a movie star, like Nancy Kwan in *The World of Suzie Wong*. She wanted to learn French and live in Paris. She was petite with a porcelain complexion, and with a distinct aquiline nose. She never went out in public unless she marked in her indelible eyebrows, the 1960s green eyeshadow that extended from her eyebrows to her eyelids and her Revlon ruby red lips. Her hair was short and permed. My mother and her sisters were always stylish, even though they grew up in Vancouver Chinatown wearing the most modern clothing of the day. In old family photo albums, I saw photos of my uncles strolling along Keefer Street, looking like a bunch of gangsters – wearing argyle plaid socks, pullovers and trench coats, with their full heads of black hair slicked back from their faces.

Three days after my mother turned 18, she married my father, a man 11 years her senior. She was the eldest of eight children - four younger brothers, and three younger sisters. Getting married was what was expected of my mother. Education took a back seat. Her mother, my grandmother Lai Fong, was widowed with eight children to bring up when my grandfather died in 1940. My mother was 13, and it was her responsibility to take care of her younger siblings. Like her sisters, my mother went to work on the farms alongside her mother, as soon as she was seven or eight and then in the fish cannery, one of the many that prospered in Vancouver at that time. Once, she bragged to me that she was kind of a head girl on the line, telling everyone else what to do. That didn't seem to be problematic for her, being how she was the firstborn. I had a hard time visualizing her stomping around in rubber boots, scaling and gutting salmon. But back in those days, you did what you had to.

My grandmother, a short pudgy woman with a stoic face, was supposedly a farm girl from Guangzhou. She shuffled around in her two-story brick Chinatown house in slippers, a long skirt, and a loose-fitting dark blouse. She was so opposite to my mother. Plain is the most fitting adjective for her. Her hands were working ones: nails short and rarely polished. She wore no makeup, except for the occasional lipstick. Like her daughters, my grandmother always had a perm in her hair. Her only adornments were the gold and jade jewelry, especially a ring on her finger and a jade heart she wore under her blouse.

I was given a jade heart by my mother. She told me that jade changes colour over time. If you were a good person, it would turn a brilliant jade green – so lustrous that the colour would take your breath away. I wore that jade heart under my clothes religiously for 10 years, hoping that it would change colour. It never did. I guess I wasn't such a good person after all.

One would think being Chinese meant having straight hair. Not my grandmother, aunts and one of my uncles, who opted to put perms in their hair. Even to this day, one aunt perms her hair. If being Chinese meant having black hair, then I want my money back. If you're going to be Chinese, you want to have black hair. My hair was never blue black, like my big sister's and my big brother's. Now that's black hair that shines blue black in the sunlight. No, I had mousy Chinese dark brown hair, like my mother. On an explanation, my mother was just messing with me that maybe, our hair colour had something to do with the water we drank. My mother was also lighter skinned than most Chinese, with freckles. I so-o wanted to be like my mom. I grew up wanting freckles like the ones on my mom's face and the fair-haired girls in my elementary school classes. I did get freckles later in life. Only to fret about then. I learned later that they're called liver spots, indicating my system was abnormal. How beauty changes in the eye of the beholder. So now that I've got them, I now conceal them with makeup.

My grandmother was a resilient woman. She adapted to her environment. She thrived. She was entrepreneurial. She made her own whiskey, labelled per batch number, which she sold discreetly in brown paper bags, during Sin City's prohibition years. Her grandson was used as a mule to the distributors. She prospered. Even after her children had moved out of her house and she babysat her grandchildren, she helped other people in the community - men and women, by feeding and clothing them. Her sons and daughters would become immigrant success stories. Several of her grandchildren would be the first to go to university and one of them, me, would set precedent.

My memories of being babysat by my grandmother are indelible in my mind. From my warped toddler's perspective, my grandmother was this larger than life troll. In reality, she wasn't even five feet tall. She didn't speak English. I didn't speak *Taishanese* (a Chinese dialect of Cantonese). When she spoke at me, hoping that I would learn

Taishanese by osmosis, she scared the living daylights out of me. In the end, I learned best to stay out of her way. I would spend my day running around in circles in the front parlour. If I was completely bored, I'd pick apart the smaller fake flowers in the vase that graced a round table in the room that separated the front parlour from the dining area. There were so many flowers. I'm pretty sure she never noticed any of them were altered.

When lunch came, my grandmother set a bowl of dumpling soup in front of me. Her dumplings were so big that they were hard to chew on and stuck to my mouth when I tried swallowing. I didn't like their bland taste or gelatinous texture, but I also didn't enjoy my grandmother's death stare. I swallowed them as quickly as I could without choking. It's no wonder I got constipated under her tutelage.

It is true that a child's early childhood will have a profound impact on his or her well-being as an adult. My mother and my father fought constantly. Behind closed doors, the abuse and violence prevailed. My parents were two wounded souls who had only one mission in life – survival. Their union was doomed from the start. My father ran away as a teen to get away from his stepfather who beat him up.

My Uncle Victor told me decades later that my mother was only married two days when she came home to tell my grandmother that she was being beaten. My parents lived across the garage in a shack – spitting distance from the family house. It was an arranged marriage. There was no such thing as divorce. This would be a loss of face, considering my grandmother threw a fancy wedding party in the upscale Hotel Vancouver. My mother, in her ruby red lipstick, spiked heels and dark coloured silk *chemsang* – the skin tight high collared dress with a slit up the side, posed leaning against some stairs, looking like a Chinese Hollywood starlet. Her fantasy of fame and fortune was her escape from her life lacking love, support, and recognition.

My uncle claimed he went over to see my father and raised hell. He would beat the crap out of my father, if he ever laid his hand on my mother again. Over six feet tall and bigger than most Asians, my Uncle Victor could have easily killed my father. On second thought, he backed off. My grandmother told my mother that she had to suck it up. A firstborn daughter into a Chinese family wasn't exactly revered. A son is much to be celebrated about. A firstborn daughter is like a factory reject. In my mother's case, there was no place to send her back to.

Like my mother, my grandmother had had an arranged marriage to a man more than 10 years her senior. Nobody in the family will deny or confirm if he liked hard liquor, smoked cigarettes or spent many hours in the back shady rooms playing *mah-jong*.

My mother was an Asian Cinderella trapped in a prison, all pretty clothes stashed away in the closet, but no ball to go to. She quipped at me once that she wanted to marry another neighbourhood boy. He became a very successful pig farmer and raised three nice boys – all who graduated from University of British Columbia (UBC) in engineering. Later I learned after she died, my mother told her best girlfriend that her relatives thought since she was so pretty, she could have had a stellar career waiting on tables and pleasuring men. After all, prostitution was no big deal. My mother told us that our father's mother ran a brothel on Yates Street in Victoria, when she first came to Canada, before becoming a dressmaker. It was obvious that my mother wasn't fibbing. None of my father's two younger brothers looked related at all.

While tweens at my school romanticized their future of fairy tales with handsome rich men to marry them and making babies, I flatly told them that I wasn't getting married. I wasn't going to have chil-dren. I also didn't believe in unicorns and pigs could fly. For sure, I told myself that I wasn't going to end up like my mother.

Chapter 6 –
Nerd Girl

My father's idea of a good time was to dress up in our Sunday finest (even though we rarely attended St. Chad's Anglican Church), then indoctrinate us by parading us through the University of British Columbia Campus. When Simon Fraser University was opened, that was our second official family field trip, other than extended family dinners at Grandmother's house on Keefer Street – which was best described as dine and dash.

School was my escape from the temple of domestic violence. In my mind, I justified that it could be worse. Way worse. We weren't living in the east side of Vancouver or in Chinatown. My father wasn't an alcoholic. My mother wasn't on anti-depressants (or maybe she was). I wasn't molested by an uncle. There wasn't enough food on the dinner table. My parents struggled to make ends meet. I knew that. While my life at home was a sham, my school life made me a nerd girl.

I loved school. Come elementary school, the teachers treated me as a novelty, as we were the only visible minority students at Trafalgar Elementary School, situated on the west side of Vancouver, as far away from Chinatown, as possible. Ironically, there was more diversity amongst the mathematics teachers than my classmates

– Jamaican-born Mr. Rogers taught Grade 4, Miss Valentine taught Grade 5 and East Indian born Mr. Gobi taught Grade 6. Mr. Gobi held us captive with a banner above the chalkboard in bold black capital letters: "Procrastination is a thief of time."

Mr. Schmidt, my Grade 7 science teacher, a slouching middle-aged man with a broken nose, challenged us to enter the B.C. Youth Science Fair with the caveat that the school would supply chemicals and lab equipment. I didn't care what the other students thought. Not being white was my alibi. That year, I bagged a humongous silver trophy, which I didn't get to keep in my bedroom like I planned. Instead, the principal from Prince of Wales High School locked it up behind a glass cage.

I was 13 when my father died of cancer. He was 52 and was working for the federal government as some sort of paper pusher. His life was bittersweet. He died victoriously, having completed his high school diploma through correspondence, which led him to his late in life career switch. My mother was 41. My big sister was 19, big brother 15 and younger sister 11. Although I bawled my eyes out at the funeral home's mourning parlour, I was relieved. My mother was free of violence, which haunted her for many years to come. Our house, which previously had a noise ban in effect when my father studied, was lifted. However, we had all become so accustomed to this silence, that music was rarely played or the radio turned on.

I got to visit Josephine, my best friend, as often as my heart desired. We both read *Nancy Drew* books. With older brothers that ribbed her, she fancied my company. Josephine, still the tallest girl in class with curly brown hair, spoke in big words that only learned people used. She came from a big Irish Catholic family. Their house smelled of pancakes, maple syrup and bacon – white people food that I so wanted to stuff my face with. Out of my Chinese conditioning, I declined eating, out of politeness.

My mother had a very hands-off approach to parenting us. By day, she worked as a bookkeeper for a small stationary store. Her day started running a load of laundry and ended with housework. She never told us once what we could or could not do. After all, she was the only working mother on the block, now the only single one and a very independent one. Her only pleasure in life was dressing well, an easy feat, as she was tiny enough to fit into sample designer clothes being sold at bargain-basement discounts.

She never meddled in our affairs. We never asked. My mother did nag us. She repeatedly told us that when we grew up – we were to leave home and never come back again. My brother who became a landscape architect took her words verbatim.

Chapter 7 –
Underqualified

The following summer after my father died, all I could think about was making money. While my mother was at work, I scoured the newspaper want ads for work. A larger than normal sized ad, caught my attention: "MODELS WANTED" – women, any age, and any race could apply. Hours were flexible. Earning potential was excellent. For interview, call a phone number. I called. A nice lady, who only went with her first name Mary answered it and scheduled me in. I had visions of being on the cover of *Seventeen* magazine. She gave me an address to go to, and told me to go to the back door of a house. I was 14, and while I had reached puberty, nobody would have guessed it. I still looked like I was 10 or 11.

I took the two-hour bus ride, three transfers to North Vancouver. The address was a lacklustre two story house with an unruly garden. I came to the back of the house and knocked. A pleasant mid-30s long-haired woman wearing a hippie caftan greeted me at the door. Trailing behind her was the photographer, Mary's boyfriend or common-law husband – Mike. He was a tall nerdy dark-haired guy with black plastic square-rimmed oversized glasses, wearing jeans and a shirt. They took me into the studio.

It was a windowless basement room with several floodlights over a bed. They ushered me to sit down, as they explained what they were up to. I was assured that Mike was a professional photographer, and what they were doing was totally legit. But I had to trust them. I remained cool and collected, pretending now that I was really 24 and not 14. Surprisingly, they didn't ask me how old I was. They didn't really care. They were anxious to see how the camera would take to me. They took photos on speculation and if they could sell my photographs, the compensation would make my co-operation worthwhile.

So today, I wouldn't get paid. I just had to model. If their buyers liked me, they would call me back for more photos. I asked them who bought their photos. Mike said that Fortune 500 compa-nies like to use models to sell products to customers, in the same demeanor that he was about to walk into a bank and make a deposit. I complied. I was curious. Mike showed me photos they took. There were semi-clad desperate women – some with pouts, some leaning over, exposing their breasts from a V-necked T-shirt, some smiling and some more mysteriously appealing.

I was conned. I was misled by their advertisement. But it was too late now. I sighed to myself, hoping they hadn't notice. I'm sure they did. I probably looked a bit pale.

Their trophy model to date was this doe blue-eyed blonde with a pixie haircut and flawless skin. With a melancholy look on her face, she was photographed only in her panties, kneeling, shoulders back and her perfect perky breasts jutting out. Mike affirmatively, said, "This is what they want."

I twitched. Mary noticed my loss of enthusiasm and said encourag-ingly, "You got to start off somewhere."

Mike and Mary eyed each other, as they noticed I lacked the infrastruc-ture that their customers were looking for. Even if I stuffed my bra with

Kleenex, I still lacked infrastructure. They were desperate for photos. They compromised their integrity. They settled for the next 30 minutes, taking photos of me, wearing a man's shirt, with the two top buttons unopened. I felt an anti-climax. I had this sinking, real sinking feeling that I was down the wrong rabbit hole. I went along with Mary and Mike's charade. I politely thanked them for their time and quickly sped from their premises, like I had seen the kiss of death.

I boarded the bus to go home. There she was. Fully clothed in a white A-line mid-calf skirt and buttoned down short-sleeve lavender cotton blouse with a Peter Pan collar. There was Mike and Mary's protégé, sitting across from me. She was the blue-eyed blonde with the pixie haircut and nice breasts. She had the same blank look on her face in real life, as she had in the photos. I felt an overwhelming grief for her.

She was older than me. I'd say 19 or 18, but I could be wrong. I wanted to go over and say hi to her, but that would have been totally inappropriate. She was untouchable. I wondered what she was thinking when she took off her clothes that day to pose. I wondered if there was other stuff that Mike and Mary expected her to do. I wondered how she felt about what she did. Were her parents mean to her? Did she finish high school? Did she have a boyfriend? Did she work as a secretary somewhere? I couldn't have saved her from going down the path she was on. I was just as desperate as she was – looking for a get rich quick scheme.

I knew Mike and Mary weren't going to call me back. For that girl with the pixie haircut, she was probably called back again and again. Her chances of becoming a porn star looked pretty good. I could have become a porn star, too. For that, I'm pretty sure I was underqualified.

Chapter 8 –
Working Teen

I knew for a real job that I'd have to wait until I was 15. I chose not to. I was 14 when I got my first job, housekeeping for a rich lady in the snooty Shaughnessy district. I was paid $1 an hour, which was quadruple of what babysitting paid. My client was a trophy wife who spent her days lounging at the elite Arbutus Club and on the telephone gossiping. Their house, rather mansion, was gorgeous in its splendour that took my breath away. I gawked at her sorority photos that graced the hallway. She meant well, but a tad patronizing, when she gave me her clothes that shrunk in the wash. I was always welcomed to eat something out of the fridge, but it was always empty. What I liked the most about my first gig was the gleaming upright Heintzman piano in their study. When she was out and my work was done, I'd slide my hands down those ivory keys. Then I'd lock the back door and put the key under the mat, then flee down the moss-covered steps and trudge over 30 minutes to get home.

By Grade 7, my marks improved drastically. The eyeglasses that I began wearing in Grade 5 brought my world into focus. The girls in my class called me "Granny" because I was starting to make them look bad. There was another nerdy girl in my class, Heather Harris. She was a new arrival to Canada from England and spoke with a heavy accent. She was reserved and dressed differently, not poorly, just differently. We became good friends – as we were the social outcasts.

When I turned 16, I got a real job working for Peter's Ice Cream on West Broadway in Vancouver, which has morphed today into a Tim Hortons coffee shop. I gave Heather the cleaning job. I was elated to work at the ice cream parlour. There was the possibility of getting tips to augment the minimum wages we earned.

But the only girls who were getting tips were Carol and Pamela, two sisters who worked downtown as secretaries and as waitresses during the evenings and weekends. Both were single smoking women in their late 30s with no children. After their midnight shift ended, the duo bar hopped. Carol had big bleached blonde hair and mascara that dripped around her big blue eyes. She dubbed me "flighty sneakers". Carol knew when she'd need anything from the back of the restaurant, I would stop what I was doing and run back. Pamela was a freckle-faced brunette with straight bangs and pageboy hairdo. Both were well endowed and complained profusely of how their 'girls' were getting in the way of their job.

I used to wonder why two office workers would take on this minimum wage job. I figured out later the ice cream parlour was also a place where men trolled for sex. The two sisters' moonlighting job to their waitressing job of entertaining men was always done very discreetly. After all, Carol and Pamela were professionals. A wispy blonde, who wore no panties under our white polyester uniforms, which became transparent when wet, was a threat to their territory. She was fired after a week. Dirty old Paul, the ice cream parlour owner, meandered around the joint with a cigarette hanging out of the corner of his mouth. I bet that pervert was their pimp.

At that time in my life, a retail job at the department store looked like a pay upgrade. But I had no retail experience and no connections.

Oriental sales clerks were still an anomaly. Woodward's Department Stores had a wall of their high school teen representatives, and I wanted to be one of them. When I asked my mother that I wanted to start wearing makeup, she thought I was a bit premature, but went along with lending me her mascara. I did have my own money to afford some blush. Eyeliner was a waste of money. I couldn't and I still have a problem with applying that stuff with finesse.

I took off my eyeglasses and walked into Woodward's and spotted Jade, the pretty Chinese-Canadian (actually as white as me, as she spoke English without an accent) girl in the Woodward's girl red uniform. I asked her how she became a member of the Teen Fashion Council. Jade told me to call Elizabeth Taylor, not the actress, but the matronly woman who headed up Woodward's marketing. She was a fashionista with flair. Mrs. Taylor only wore black dresses that matched her pitch black sprayed hair, heavy black eyeliner, false eyelashes, and pancake makeup.

I mailed Mrs. Taylor a hand-typed letter with hand-drawn flowers explaining that I wanted to be a member of the Fashion Teen Council. I followed up with a phone call. She was enthused about having me come in. Little did I realize that this was an unpaid position. Those monthly Saturday meetings were held downtown at 8:45 a.m. I had to eat crow when I found this out, after staying up until past midnight the night before, schlepping up corned beef sandwiches and Campbell's soup for late night diners at Peter's. Then after the Fashion Teen Council meetings, I'd come home hungry and exhausted. Then I would study and then head off to Peter's for the 4:00 p.m. night shift, which I didn't argue about because more hours meant more money.

Working for Woodward's meant wearing a cherry red Fortrel 1960s uniform, which made me feel that I was a big shot. My face graced the wall of Woodward's for 1970-1971. Perhaps, there was another young girl who looked up to me. True enough, the mascara stuck to my straight eyelashes, which I didn't have patience to curl. I was not tall enough to do the cat walk on the red carpet like the two prettiest girls on the council, including Jade, who was one of Elizabeth's favourites. I did get to work in retail at Woodward's – in sewing machines. I and another shy girl got called for the unpaid Easter promotion, wearing hot pants and taking care of bunnies and feeding urinating baby goats on display. That gig ended badly. Mrs. Taylor thought this would be the cat's meow. I begged to differ. I bailed out, calling in sick with the flu. That just buried my foray into the fashion industry, then and there. I couldn't have been more pleased.

Chapter 9 –
Games Nice Girls Don't Play

On his deathbed, my father pleaded with us children to go work after high school and support our mother. I graduated from high school with straight A's, except for physical education classes. I could have gone to any university. As a middle child, I felt it was necessary to compromise between what my father wanted and what the ivory towers offered. I enrolled in a two-year food technology program at the B.C. Institute of Technology in Burnaby, B.C. – the municipality east of Vancouver.

In 1973, I graduated from the BC Institute of Technology, top of my class with my bio-sciences technology diploma, specializing in food technology. For this academic exercise, I was rewarded with a non-precious medal that sits in the bottom of a drawer somewhere. Book smart doesn't equal street smart. I was naïve to think that just because you can excel academically that you will get the best job. As a food technologist, the coveted job after graduation would be to go work in research for Kraft Foods. Our final second year project was to come up with a viable new food product. I came up with freeze-dried scrambled eggs. For sure, I thought I had earned the accolades to work there. But destiny had something else in mind for me. Sarah, the sassy flirty green-eyed redhead would land that job at Kraft, despite her mediocre marks.

The response I received from the Kraft Foods recruiter turned out less than stellar. I didn't understand. That job was mine. Kraft sent me a Dear John letter that we were breaking up and yet I never put an hour into our working relationship. The message was loud and clear that there was no future for me at Kraft Foods. The recruiter insinuated during our interview that I was an over-achiever and that I should consider going back to university. Some people said it was racist for Kraft not to hire me. I'm sure Don, the tall dark-haired mustached man in our class who had a degree in psychology before becoming a food technologist, would attribute all of our life tragedies to "improper toilet training" – the bane of our existence.

I did get a real food technology job, working for Robinson's Fine Foods in the industrial part of Burnaby. I worked in quality control. The pay was the pits. The working conditions were on par for a food processing plant. I was assigned to quality control, taking sugar and pH samples of pie fillings, or the sodium bisulphite levels – a preservative. The plant reeked of sulphur odour. The floors were slippery and wet. Coffee and lunch breaks were segregated. Buzzers dictated when you could pee or take a break. The women who worked in the plant were middle-aged women who never saw sunlight or a gym. My breaks were spent waiting in line to pee, as they just had the minimum number of cubicles. Once, I snuck out to the truck loading area to eat my bagged lunch, just to catch some sunshine.

The recruiter from Kraft Foods was right. I lasted three days at Robinson's Food. They never paid me. I never looked back. I went to work for Environment Canada in North Vancouver in their freshwater biology lab as their freshwater biological technician. Apparently, I would be no threat to anyone. I would work for biologists and engineers in a dead-end job. I wasn't going anywhere. I'm sure the pink and white gingham empire waist dress with a bow at the back, further communicated during the interview my desire to be a lady with no ambition.

Chapter 10 –
My Age of Aquarius

When the Broadway musical *Hair* was released in 1969, there was this buzz about the dawning of the Age of Aquarius – when love was in the air, as the planets were in perfect alignment with the stars. Two years later, John Lennon and Yoko Ono produced the hit *Imagine* – a poignant iconic song that sent a message around the world pleading for peace and harmony. The Age of Aquarius spread fairy dust through the cosmos.

My high school classmates said they had the time of their life in their 20s: going to parties, hitchhiking around the world, getting stoned, drinking themselves silly, and getting laid. Their Age of Aquarius struck them in the right chord, as they would 40 years after graduation speak of fond memories for their coming of age. My dawning of the Age of Aquarius was thwarted by one of my astrological planets trapped in a black hole. I still got asked for ID when I went to nightclubs. I was 25 when I got asked by a 15-year old boy at the ice rink to skate around with him once holding his arm, to impress his peers heckling him in the bleachers. I agreed out of pity. Perhaps, there is something called karma. The repercussions of a random act of kindness could manifest itself, at a later date.

If only I could have been more normal. Why couldn't I have been satisfied with what I already accomplished? Why didn't I agree to spend the rest of my life working for Environment Canada in West Vancouver; counting micro-organisms, like many of my colleagues at that time did? I would have done okay with my life. At least, I had left the pink collar ghetto and earned more than minimum wage.

As a freshwater biological technician, I peered down a stereoscope in search of *Daphnia*, transparent water fleas, about the size of a grain of sugar. When the water was clean, the water fleas thrived. When Weyerhauser Pulp Mill had to relieve itself into Kamloops Lake, the fleas found themselves on death row. Then algal blooms infiltrated the lake, upsetting the ecosystem, sometimes, killing fish due to oxygen deprivation. In 1973, it was legit for Weyerhauser to dump effluent into Kamloops Lake, as back then, "the solution to pollution was dilution."

Environment Canada monitored the lake diligently. When the ecosystem looked like it was going psycho, an environmental officer told Weyerhauser to push back on dumping, until a better day.

I reported to Louise, an attractive tall woman. The worst part of my job was to endure blatant sexual harassment by another biologist, a swarthy shifty-eyed Catholic man. Quite frankly, strapped with a stay-at-home wife of increasing weight and four school-aged children, Bert was in a rut and wanted more drama in his life. He openly lusted for Louise, like a pit bull in heat. While peering over her microscope, Bert would grab and kiss her. Louise fought back, but never filed a complaint to human resources. This was 1973, when men were men and if a woman didn't like the way she was treated, she had two choices – endure it or quit.

We were low on the food chain. Bert had less work experience and poorer university grades than Louise. Yet, within a few months, he was promoted to a more supervisory position and got something that made Louise envious – field work. The argument was that Louise wasn't strong enough to pull the puny one litre sample bottles a few metres below the lake's surface up to the boat. Even a 70-pound kid could do that.

Louise wouldn't take crap. She was assertive, which in those days was not considered an attractive quality for a woman. Higher up the food chain, our pimps were engineers. Louise and I didn't consider them to be the brightest people in the universe. When they came to visit us from the mothership, the downtown Vancouver office, we would act like everything was hunky dory. We were so docile. They could have probably tortured us, and we wouldn't have said a peep. Then Louise finally convinced Bob Kingston, who I believed did dick all, to go to the field. Of course, there was concern that they couldn't send a single woman to the field. I got to be Louise's Sherpa. We got our token field trip to Kamloops. We came back to the lab and nothing changed. Bert got to dissect fish guts and identify the insects the finned critters were chomping on. Both Louise and I carried on with the zooplankton and water flea counts.

Within a month of working for Environment Canada, I realized I wanted more. Louise's father convinced me that I could be an engineer. Mr. Thompson was a successful civil engineer, who ran his own consulting company. He lived in a beautiful waterfront West Vancouver home, married a wonderful woman and was father of three children. Louise's family was what I yearned to have been born into. Once, after work, Louise brought me to her parent's place for dinner. We had barbeque chicken, corn on the cob, roast potatoes and green salad with Kraft's French dressing – the sweet tangy dressing tantalized the tips of my taste buds and affirmed what I had fantasized growing up – white people's food tasted good.

Maybe, it was just a mirage, but there was some speculation here that engineering was in my arm's reach. Louise's father had enlightened me in this aspect. I worked full-time and went to night school at Langara College to complete first year mathematics to get into engineering school. I wasn't alone. There I met two mature students who were going to study engineering, too. There was Jack Tremblay, a long-haired pothead in his late 20s, trying to seek his father's approval; and Ron, a married mid-30s draftsperson looking for a career upgrade.

My mother was amused by my decision. She did remind me that two gals from my high school, who also attended BCIT, were now established in their prospective careers – one in business administration and the other in nursing. Both gals were gainfully employed in their respective professions, drove their own cars and flaunted engagement rings – all before the womanly age of 21. Maybe, because my mother was a single mother or because I felt I wasn't pretty enough, I never believed that getting married and having children was in my cards.

Chapter 11 –
They Told Me There Could Be Rape

On the morning of registration, I woke up with hope. I was obviously dreaming. Déjà vu, as they said in the 1970s. I was already in engineering school, and I felt fine. I mean I felt really fine. I am pretty sure that I wasn't high or stoned. At that pivotal moment in time, I knew I could tear up my acceptance letter to Forestry Science, which was my second choice. I could have ended up being a tree hugger. I liked plants, and I especially liked trees. My future was green, working as a researcher in a provincial greenhouse, concocting ways to coax the trees on becoming bigger and stronger faster.

Not today, I told myself. I was going to be an engineer … even when some girls heard that I was going to study engineering, they barked back, "I wouldn't do that. You'd get raped."

It was too late now. It was September 1974. I stood in line for engineering school, clutching my coveted acceptance letter. I was nervous and self-conscious – as I was somewhat a hypochondriac back then. The night before, I obsessed what to wear. If I wore a dress and heels, then would they think I was just looking for a man? If I dressed unattractively, then they'd think I was a lesbian. There was a fine line here on being attractive vs. unattractive. Finally, I settled on my hand-made tailored two-piece brushed cotton pantsuit, in air force blue – a colour that strangers remarked was becoming on my Oriental skin. I wore the only pair of walking shoes I had – Clarks' crepe-soled tan Wallabees.

There were over three hundred persons standing in line, mostly casually dressed young men – some jocks in baseball jackets, a few long-haired Hippies in tie-dyed shirts, small town boys dressed for a job interview or Sunday school and the occasional dude who arrived from a disco dance in a paisley polyester knit shirt. Small talk was limited to those who came from the same high school.

The eyes of my new peers darted around aimlessly. I could feel them staring smugly at me, possibly undressing me with their eyes, to kill time. The line moved at a slug's pace. There were no cellphones, Internet, laptop computers, Facebook or social media. My only comfort that day was it wasn`t raining. It was a quintessential fall day. The slight chill in early morning warmed up to ambient room temperature. The effervescent blue clear sky was marred by the occasional cumulous cloud. Ivy vines climbed the walls of the stone buildings, with tendrils scrambling for the top. Campus tree leaves only sparingly transitioned from apple green to yellow.

Registration was done the good old-fashioned way. A female administrator sitting at a desk with a graphite leaded pencil ticked our names from a list to confirm that we had paid our first term in advance. Amongst these young males, I felt like a dildo being sold in the children's toys section of Woolco (or Walmart nowadays), as my eyes fell short of any other girls in first year.

***

Four months earlier, I had stepped into the Dean's office to pick up an engineering registration form. In the musty oak paneled room, there was a solid wood counter that kept anyone from running into the Dean's office hidden behind a frosted glazed window door. Mildred Kastner, a menopausal spinster with a raspy voice, sneered at me over her bifocal glasses. Over the ringing black rotary phones and the banging Remington typewriters, she bellowed at me, "The job has been taken."

I was confused, "What job?"

She replied, "The junior secretary. We hired her last week."

I said, "But I'm here to pick up an engineering enrollment form."

Mildred kept her poker face, efficiently grabbed an application form and plunked it down in front of my nose, "Do you have a 60 per cent average?"

"Yes," I replied.

Mildred demanded, "What about a 65 per cent grade in mathematics?"

"Yes," I replied.

Avoiding eye contact, she cleared her throat, "Here you go."

"Are there other girls in first year?" I peeped back.

She shrugged her shoulders and cackled, "A few, I suppose."

To say Mildred was intimidating was an understatement. We thought the Dean used her as his shield from pestering students so he had more time for writing technical papers and important presentations that would embellish his academic curriculum vitae. She left such a mark in our minds, that forty years later, the class of 1978 would discuss setting up some recognition in her honour.

***

The minutes in line seemed like eternity. This mundane existence was charred by a pack of boisterous senior engineering students sporting the infamous UBC Engineering Undergraduate Society (EUS) preppy red and white cardigans, blowing on kazoos and chanting off-key,

"We are, we are, we are, we are, we are the engineers.
We can, we can, we can, we can, we can demolish forty beers,
Drink rum, drink rum, drink rum all day, and come along
with us,
We don't give a damn for any old man who doesn't give a
damn about us."

Trailing behind were two students, one carrying a clipboard. As if they discovered plutonium, they spotted me cowering, singling me out, "There's another one!"

The stench of cheap beer followed them. The guys ahead of me turned around, gawked at me, and then turned their faces forward. My ears were burning red. I wallowed in discomfort. Then from above – the roof of the civil engineering building – water filled balloons were hurled at us, exploding upon impact. Profanities followed. The orderly line dispersed. My left shoulder was soaked. I calmly took off my jacket, disgusted and told myself, "Now what have I done?"

That pep talk I gave myself three hours earlier that morning just lost its currency.

Chapter 12 –
The House of Nerds

Many years later, I was dwelling on the meaning of life and googled the terms "nerd" and "geek." *Merriam-Webster* defines a nerd as a "contemptuous dull person or bore" whereas, a geek is "an intellectual especially in the field of technology who is disliked" or, the pre-technology era – "a carnival performer often billed as a wild man whose act usually includes biting the head off a live chicken or snake." Then I figured this out. A nerd can be a geek but a geek never a nerd. Think about it, my friends. One thing is certain, neither nerds nor geeks win Miss Congeniality contests.

My university peers grew up on Meccano, Lego and Tinker toys, read *Scientific American* or *Popular Mechanics* and got high from building model airplanes. Their high school guidance counsellor observed that with their nerdy outlook, and disdain for stringing words together to compose a coherent sentence, they best consider a career in engineering. That's what boys did in the 1970s. It was a logical thing to do, especially if their father or uncle or big brother studied engineering. There's no need to question the guidance counsellor's advice as engineers aspire for a steady paycheque, marry a nurse or teacher who will quit her job to stay home as soon as their firstborn comes along, buy a house in a nice middle-class neighbourhood and drive a reliable Volvo.

The nerd is hardly a threat to the status quo. In the case of engineers, they *are* the status quo. No matter what culture you come from. Even in 2018. I met a 26-year-old East Indian woman stressed over the dilemma that her parents want to set her up with a 31-year-old East Indian boy from out of town. Their biggest selling point is that he's an engineer. How can she refuse such an offer? Well, except, she already has a boyfriend she loves. The nerd may not be suave and debonair, an assassin or the next prime minister, but when tradition persists, he is good marriage material. Even so, the nerd mystique can be a bit daunting.

I stumbled into my first year engineering class – mathematics – wearing the same air force blue two piece hand-made suit (but now dry). It was a long tiered lecture hall, each row was a couple steps up from the front row before. I looked up. My inner voice squawked, "Oh shit! What have I done?"

I know other boy crazy women would have rejoiced with this sausage fest. I stood there petrified, momentarily, and then grabbed the first seat in the front row, closest to the door – this was my contingency plan, just in case I would hyper-ventilate and chicken out. My eyes were riveted on the blackboard at all times. So were the eyes of the skinny blue-eyed curly haired blonde who sat beside me, motionless. The air was so still. All I could hear is my heart palpitate. My breath was shallow, my stomach queasy. A long-haired female student entered the classroom, sat down, momentarily. She checked her timetable and noted she was in the wrong room. Her abrupt departure was marked by catcalls from the back row peanut gallery. Her face turned a much darker shade of red.

There was only one other time in my life I felt this way, such dread, albeit. It was Grade 5 recess break. Marg, myself and Harriet were squealing and chasing each other around Mr. McDougall's classroom. We weren't even supposed to be there. We were supposed to be outside. Mrs. Kelso, a large middle-aged woman with messy curly greying-dark hair and disturbing hair on her upper lip was patrolling the hallways. We heard her heavy footsteps approach the classroom door. All three of us ran for cover. Marg sought refuge behind some coats in the cloakroom at the front of the classroom. Harriet was standing in the back of the classroom's corner out of view of the window in the door. I was curled up like a rabbit under the desk. We heard Mrs. Kelso grab the doorknob. She scanned the room, saw nothing, released her grip and walked away. For sure, I swear I was going to go into cardiac arrest. Today, sitting in this first year engineering math class, I dreaded being exposed as a thief who coincidentally slipped a bottle of pearly pink nail polish into my pocket as I left a drugstore.

I wasn't imagining anything when I turned my head ever so slightly to my left. I presumed nobody was looking. From the corner of my left eye, I registered another girl. I was elated. But I could barely get any facial recognition. Her head was submerged into a softcover paperback – likely an escape of our new reality sitting in a classroom of nerds.

Each day felt like the first day, for the next 30 days. I am not being dramatic here. There was this eerie silence before classes started, wondering if and when the next stunt by senior students would de-rail us. Each day that rolled by incident-free was a load off my shoulders. Just when we thought we were safe; the axe would drop. Water balloons were commonplace. High-pressured water from the fire hose would ruin our handwritten notes. Wearing nice clothes to school wasn't even a consideration. Dabs of my favourite YSL cologne, *Rive Gauche*, was essential to mask the body odour of my classmates living in dormitories with showering restrictions.

Rumour had it that engineering students or "gears" hid in the stalls of the girls' washroom by the Engineering Undergraduate Society office. I maintained a one hundred metre clearance, at all times; opting to pee in the newly-opened geological sciences building or the agricultural sciences building. I certainly was not going to get caught with my pants down.

Weeks later, the other girl in engineering math and I mustered enough courage to give each other eye contact. Finally, she motioned me to sit in the empty seat beside her. That was just an opportunity to be heckled by the guys in the back row, "What two girls in class are sitting together?" This was followed by more heckling, "Are you two lesbians?" We glared back, as if we had the power to tell the Federation to take over the Klingons.

There's a cliché about being careful with the company you keep. Nerdocracy was inevitable. While other friends in less demanding faculties were out partying, we were studying, except for the odd mutant kid (the one with Asperger's or something) who learned everything through osmosis. In no time, I owned three slide rules, (since the PC didn't arrive until many years later). I had a six-inch plastic one that fit nicely in one of the vinyl plastic pocket protectors (if you were a guy) or in my purse. Then I had a 12-inch plastic slide rule that came in a light grey hard plastic case and a 12-inch bamboo more eco-friendly slide rule in a red and beige cardboard box that was superior to the others for not only its tensile strength but its ability to slide with ease.

I was no match for some of my peers, especially, the electrical engineering students, who prized themselves slightly a cut above the rest of us. In 1974, Texas Instruments released a handheld scientific calculator SR-50, with the intention of replacing slide rules, which still gave you answers to third decimal point accuracy. Wearing a SR-50, especially, one of the latter models with mini-programming cards in a holster that slipped through your belt, was the epitome of geek chic. Cash flow seemed to be a deterrent for the vast majority of us.

I resisted early adoption as my peers complained that battery life was as short as the male orgasm.

Resistance turned out to be futile. In no time, the vast majority of my peers had some sort of handheld calculator slung around their hips. I took the hour long bus ride to some industrial warehouse in Burnaby. I rang the bell on the counter and gave Joe, a sheepish looking balding man in wire-rimmed glasses and a shirt one size too small, $160 cash. That was one month's rent for a Melcor calculator, one of the earliest calculators. The Melcor had two storage memory registers and could do all the basic calculations – like add, subtract, divide, and multiply. My beloved Melcor had a sudden and shocking death. It slipped out of my back pack and hit the ground, smashed to smithereens. The timing couldn't have been more perfect. I just finished my last and final fourth year exam.

Chapter 13 –
The F-Word

It was a shock to the UBC Engineering Undergraduate Society that the first year class of 1974 had 14 women. Up to then, first year was 100 per cent male or tainted by one or two females; with their graduation highly speculative. In the quagmire of the upcoming International Women's Year 1975, the senior students had their minds made up before we walked into the classrooms that we must be feminists. Hand-made banners ripped from a roll of white paper graced the halls messaging, "Feminists – Go Home" or "Down with Feminists."

The f-word, i.e. feminist, was never mentioned in passing amongst us women engineering students. It had such a negative connotation that fear struck a chord in me. That would make me a castrating man hater, which I don't think I was. I took all the obligatory high school home economics classes, and was highly proficient at cooking, baking, and sewing. I was always hyper vigilant about the way I presented myself. I was told that I was not just to think of myself, but for the other women who follow behind me. I usually wore slacks to class. Once a week and never on the same day, I'd up the ante. I'd deliberately wear my red ruched top with a knee length brown Fortrel knit A-line skirt to class. The guys did notice me dressing up, but chivalry was not to be expected. I kind of crossed the line when I entered engineering.

A year earlier, I had gone to Simon Fraser University for their free aptitude and psychology tests. They ascertained that I should be a musician, economist, movie producer, or doctor. Medicine was a consideration. During first year engineering, I volunteered Saturday evenings in the emergency ward at the Vancouver General Hospital for a year. If I was going to be a doctor, I better get used to the stench, the misery, and sickness. I didn't have to watch *General Hospital, Marcus Welby, EMERGENCY*, or any of the contemporary doctor shows back then; I was living it.

The emergency room always gagged me in my gut. For the trauma I witnessed from 7:00 to 10:00 p.m. on Saturday night, I needed a week to recover. Saturday nights brought the troubled and marginalized to emergency – the regulars, the drunks who were picked off the streets flat face in their vomit but more heart-wrenching to me were the young women who had legal abortions but developed complications – in over 90 per cent of the cases unaccompanied by their former lovers. The most memorable botched abortion case was a little Asian girl, no more than 14. With her short hair, I thought she was a little boy. Hysteria hung around this teen. Her boyfriend and three generations of family members from both sides were screaming upon her grand entrance. Whispers spread throughout the staff that she was the outcome of the proverbial knitting needle abortion. One orderly told me she was hemorrhaging. Her worst-case prognosis would be an emergency hysterectomy.

The drama in the hospital had me running back and forth between the emergency room and pharmacy to pick up drugs or medical supplies, fetch bedpans, grab drinks for incoming patients, hold doors or reassure patients that the doctor was coming. Orderlies, doctors, patients, and nurses all engaged with me. There were so many patients in crisis. One unlikely culprit was a well-coiffed country club wife with sparkling diamond stud earrings, like the ones Catherine Deneuve always wore in the 1970s N° 5 CHANEL ads. Approaching menopause, her four children had grown up and left home. Her husband, a successful businessman spent time schmoozing and playing golf. She had everything that any woman would want, but she claimed that she wasn't feeling well. Her blood pressure was fine. Her body temperature was normal. Her pulse and heart rate were good. She was obviously lonely and needed a hobby or something. I thought to myself that maybe she should go for therapy. I couldn't give her any advice because I certainly was no M.D. After we chatted for about 20 minutes, I asked the orderly, "What will happen with her?"

He mentioned that since there's nothing really wrong with her, the doctor will likely prescribe her an anti-depressant or sleeping pill and send her home. Wait a minute. What if she's not really depressed or crazy at all? What if her husband is really screwing his secretary?

If my prognosis for this woman was out to lunch, what about the guys in my class confusing feminism with a women's state of being braless? The structured pointed nipple cotton bras of the 1960s had yielded to polyester filled structured bras and to coincide with International Women's Year, WonderBra released a revolutionary front closing bra – *Dici* or nothing. *Dici's* claims were understated as the flimsy material of the cups did nothing to cover up the nipples and yes, one's breasts would flap around like a pair of bird wings, just as the television ad portrayed.

Even though I clearly failed the pencil test (when one's sagging boobs can hold a pencil in place underneath, a bra is called for), I insisted on wearing a bra. My bra and I were inseparable. The bra shielded my nosy nipples from seeing too much of the external world. The extra layer of fiberfill cloth provided warmth to my ever so slightly protruding chest. Besides, wearing the bra was my form of misleading advertising, faking curves that I didn't really have.

In the hallway once, my classmates noticed that I wasn't braless. Two hollered, "Why are you wearing a bra?"

I paused, turned around, "What?"

One of them, the scrawny unshaven one, probably with the bad breath said, "Yeah, aren't you a feminist?"

Perhaps there was some subliminal meaning to the time when senior students slung pig shit at us during first year. Obviously, there was something Freudian that slipped by me that I didn't fully comprehend.

Chapter 14 –
I Could Never Be One of the Boys

The first time I drank a cheap Canadian beer (I swear it was a Molson's but it could have been a Labatt's) with the "gears", I came home and broke out in hives. Then and there, I realized I probably shouldn't ever consider trying out for the 40 beer club – down 40 beers in 24 hours gets you literally a crested badge of honour to wear on your EUS sweater. After all, what's the point in starting something that you can't be good at? For my body size, one small glass of wine already makes me illegal.

In addition to not peeing standing up, I knew I could never be one of the boys. I didn't intend to cause such a kerfuffle. Or go down in history. Wikipedia on the history of women in engineering describes the decades from the 1950s to 1970s as the era of "resistance to co-education in engineering schools". Student culture reflected traditions that required no explanation. The gears were pragmatic, organized, and creative.

In addition to the regular chanting of the "Engineer's Song", there was a weekly *Red Rag*, the one page news bulletin mimeographed on pink paper, which was distributed during a common class on Wednesday mornings at 9:00 a.m. The *Red Rag* was handed out in the same manner collection envelopes are passed out amongst the pews at church. Events that the engineers were proud of included their charity leg auction, chariot races, and sports activities. Hookers and strippers entertained the young men in the classroom. These were the 1970s. It wasn't a question of right or wrong. It was only a reflection of what was prevailing social behaviour. The stockbrokers portrayed in the movie *The Wolf of Wall Street* (released in 2013) were certainly no exception.

The naked lady on the horse, Lady Godiva, was an annual tradition during Engineering Week, which occurred during the third week of March. Sad but true, but renting the horse often cost the engineers more than the arts student who volunteered to ride a horse buff naked. I personally had no problems with Lady Godiva. Her legendary qualities made her rather redeeming. In decades to come, Lady Godiva would gradually be asked to cover her ass with at least a G-string and then leading up to a cape, until full abolishment from the campus.

"Blue movies" were shown every Wednesday during the mandatory mid-week hump break. The girls of our year never attended. I understand elsewhere in Canada, in the years to follow, female students watched porn alongside the guys in their class, without batting an eyelid. Strippers and happy hookers performed acrobatics fully naked beside the lecterns in the same classroom, when an hour earlier, some professor was finishing up a lecture on thermodynamics.

I give my peers credit. As engineering undergrads, they took all risks, sometimes breaking the law to pull stunts, like placing a Volkswagen Beetle several hundred feet above air on top of the university clock tower or more impressively, hang one under the First Narrows Bridge in Vancouver. Toilet paper wrapped around trees, lamp posts or freestanding sculptures was par for the course.

Situated in Vancouver, dubbed in the 1920s as "Sin City", UBC led the nation in bragging rights as having one of Canada's rowdiest engineering schools during the 1970s. In my opinion, this was due to easy access to prostitutes, marijuana, and other sins.

Akin to American colleges, the engineers had their own hazing rituals, all in the name of bonding. Tanking involved grabbing a peer fully or partially clothed and throwing him into a pool. The civil engineering department took tanking very seriously and with no exception of gender, all classmates had their turn. Eventually, female students would get tanked, beginning with the girls who volunteered on the Engineering Undergraduate Society.

The *Red Rag* and annual yearbook *Slipstick* received support of the local business community. Legitimate businesses would pay for advertisements in the year book, which posted everybody's mugshot year by year. The *Slipstick* recorded with precision – photos of all events: the Engineer's Ball, special visits from happy hookers, whipped cream body wrestling, rows of guys mooning their disapproval at being told what to do, close-ups of Lady Godiva's breasts, and whatever went on at the Smoker.

Wikipedia, coined the term Smoker as "a bachelor's party in the 1970s". It is with admiration I give the engineers credit for their months of preparation and the almost 100 per cent participation rates of the guys in the class. The tradition of the Smoker originated in the 1960s and continued into the 1970s and 1980s. In the March 4, 1982 edition of UBC's campus newspaper *The Ubyssey*, it was noted that the racist homophobic sexist *Red Rag* would be discontinued but the live sex show known as Smoker would continue to exist. Its execution across Canada varied provincially and from city to city. As one of my mentors told me, good Catholic boys from rural Saskatchewan had their beloved Lady Godiva at the University of Saskatchewan, but Smoker did not occur.

Hearsay had it that spectators included some of the professors, practicing engineers and even men in blue. The engineers took great pains in concealing the location of the off campus event, which would not be revealed to the students until the day of the event. Attendees would rendezvous at a downtown location first. No one was allowed to wear any of their trademark red sweaters, or draw attention to their presence as what they were up to. Tickets were sold for a mere $5 to $10, under the premise of serious technical topics, such as, "Lubrication Suction Pump Conference". The locations varied yearly, often outside of Vancouver's city limits in hotels in Burnaby on the east side of town. One year, it was held at a local ethnic Community Centre.

The Smoker was the ultimate bachelor party that would force everyone into unsworn secrecy, until death do they depart. A tall mild-mannered British born engineering student two years my senior miraculously excused himself from participating in such shenanigans without harassment, "I never wanted to be at an engineering function in the future with my wife, when some guy comes up to me and asks me, if I was the guy at the Smoker ... then explain to her what happened."

One introverted acne-prone kid caved in to attending one during third year, "If I don't go, they're going to think I'm a homo."

His sentiments weren't singled out. One participant described the Smoker as one big "Fuckfest". After the strippers peeled away, the hookers flooded the stage. It was common knowledge that the guy who won the door prize slipped away into the back room with a hooker for the evening. The masters of ceremony would call up the runner-ups onto the stage to bang the hookers, while he had two or three slithering over his body. Photos in the archived year books accurately recorded such illicit activities – line-ups of guys with their backs to the camera and their pants down waiting for their turn, actively engaged in various sexual positions – slightly cross-hatched so the participants cannot be identified. Despite being pissed to their gills, most of the testosterone driven males were able to perform, but the non-performers were publicly humiliated with boos and other derogatory remarks. It was a night that a few would rather forget, but due their circumstances, they would be promptly written up in the following Wednesday in the *Red Rag*, "He didn't get it up".

None of the girls in our year were invited to attend the Smoker. But in the years to follow, I heard "the more open-minded girls in engineering" attended and also went on stage too – rubbing their naked bodies with the prostitutes, as "you forget, nothing turns the guys on more than two girls making out."

Chapter 15 –
Their Terms of Endearment

Deep down, the guys in my class had a special affection for each girl in our class. But I could be wrong. Emotional intelligence was never considered part of the psyche until the twenty-first century. It was such a big deal to have so many girls in first year engineering at UBC that the guys thought it was best to recognize each and every one of us, in unique, creative and memorable ways in the *Red Rag*.

Finally, my claim to fame in the *Red Rag* would appear in second year. I was the last female to be portrayed as some sort of sex trade worker. I grimaced at the fact that my peers thought that I could make extra coin giving guys hand jobs, as I was good at shifting gears in a car. I had this sinking feeling in my stomach that there was probably nothing I could do except grin and bear it.

I showed the article to May Lau, a fourth year metallurgical engineering student. She was outraged at the ad and saw an opportunity to get my third year tuition paid for. We went to the Faculty of Law's students for their advice. As the *Red Rag* was a spoof on daily campus living and not be construed as taken seriously, there was nothing they could do. However, an apology was called for. May and I toddled over to the Engineering Undergraduate Students' Office. A puny pathetic looking mousy dude emerged. He was not at all what

I expected, not the robust testosterone divine male that engineers portray themselves to be.

Some would think that studying engineering would make us girls popular in the dating arena. Nothing could be further than the truth in 1974. We were more like sacred cows. One male student started dating another female student to the disapproval of the *Red Rag's* editorial volunteers. The two were harassed to the point that both dropped out, with the male student choosing to get a liberal arts degree instead.

During first year, for sure, we girls thought we'd all have dates to go to the annual Engineers' Ball – the hiatus of Engineering Week. I repeat we were sacred cows. I had my friends set up another girl in engineering with a date from the commerce faculty. I declined going to the Engineer's Ball first year, as I didn't have a suitable dress to wear. It was revenge of the nerds and senior prom all over again – except the stakes are higher as each engineering department competes for the best engineering display. For one day of the year, the gears preen and groom themselves like they're ready to walk the red carpet with boutonnieres in their tuxedos and corsages for their dates. While some guys hired escorts for their dates, I asked an emotionally unavailable graduate student to the ball during second year. I skipped going to third year, because I couldn't wear the same dress two years in a row, could I? For fourth year, I broke tradition and came by myself. Guess what? Nobody really gave a damn.

Chapter 16 –
Spikes at Her Elbow

Having learned from my BCIT days how tough it was to find work, I began my job search after Christmas during first year. Figuring the competition within the city limits would be life threatening, I decided to look out of town. I think it was because of International Women's Year or beginner's luck. I received a call from Kerr Priestman & Associates, a respectable Victoria, B.C.-based engineering company. They informed me that they would like to interview me for a summer position. I was perplexed how I would navigate my course schedule to take a ferry over for a day. To my delight, they arranged to interview me on campus and made me an offer, slam dunk. I was feeling pretty good about getting my first engineering summer job.

A civil engineering senior who was friends of a high school girlfriend, knew a guy who rented out rooms in his Oak Bay house to help pay for his mortgage. The rent was affordable. It was a bit of drag for me to get to work downtown – an hour commute, each way. We were the last stop at the end of the line. Rosemary, a petite brunette from Washington State, who I met at the bus stop, would make sure that I got to work on time. She would stall the bus driver whenever I was seen racing across the grass field in Mary Jane wedge heels. After calling off her engagement, Rosemary, a liberal arts major, opted to travel and take a summer secretarial job in Victoria. Rosemary was a girlie girl, who aired her sexy lacy black brassieres on the clothesline, and vowed celibacy until marriage. A year later, Rosemary mailed me a letter to tell me that she was running off to marry a man of God (a church leader, I vaguely recall). I will assume she lived happily ever after.

I couldn't have asked for a more perfect engineering job for a first-year student. I worked on a wide variety of projects. I was called to enter manholes connected to sewers in service, to confirm that the contractors installed the right diameter pipe. Other times, I was asked to hold the survey rod, when things were still measured non-digitally. When it was slow in the office, I ended up painting the office walls that 1970s shade of ochre. I was treated well in the office, by the men. They treated me like one of the guys, as best as they could. Some were surprised to meet me in person – a soft spoken diminutive girl back then who came to work in dresses and two-piece suits. One engineer at work remarked, "I always thought women engineers were tall, had short hair and wore Cossack boots."

Years later, another engineer said to me, "Yeah, you'd think they (women engineers) were so tough that they had spikes at their elbows."

Lunch times were always tricky for me. I felt like an outsider. I was drawn to socialize with the draft persons instead, just reflecting that I came from a lower working class background. Fraternizing with the secretaries would have been possibly, a professional faux pas. About once a month, our group walked down to the pier to dine at a pub where we literally sat on padded re-cycled covered toilet seats. I shunned ordering something that I would have to pick up and eat with my hands, preferring a soup and salad. I was still self-conscious about eating in public but slowly becoming acclimatized.

With the exception of Stephanie, a tall slender single blonde approaching 30, work was relatively uneventful. I could feel the tension between Stephanie and me because I was helping out Allan, a tall dark handsome man, who even my mother was smitten with. About once a month, I came to Vancouver for the weekend. Allan would show up at the doorsteps, as if he was the guy in Carly Simon's song *You're so Vain* – about to board a yacht in his grey flannel pants, neutral coloured turtleneck and navy blazer. We carpooled over to the ferry. It would have been tempting to get romantically involved with Allan. However, I had one rule – not to date the men from the office. Besides, when did I have time?

Chapter 17 – #MeToo –
Men Who Behave Badly
Rarely Make History

I don't think there were any complaints about my work performance at Kerr Priestman. However, to my disappointment, I did not get called back to work there for the next summer. I had to find work elsewhere. It was during second year that I encountered what would be the first of many trials and tribulations, that plagued me throughout my engineering career, until a new generation of men would be indifferent to gender or sexual orientation.

I applied for a summer engineering job at a prominent civil engineering firm, based in the east side of Burnaby, almost close to New Westminster, B.C. It was a ninety-minute bus ride from where we lived. Previously, I was instructed at BCIT by a senior instructor that women were advised to sign any business correspondence with their initials only. He openly admitted that industry of any kind was not receptive with women on staff. I only signed my letters of application with my first name initial and last name. My next mistake was a false expectation that since Mr. Joe Smith was an executive of the BC Engineering Association that he would be more progressive towards women in engineering. The secretary called me for an appointment. At the last moment, she asked me to bring my engineering drawings for the interview.

It was one of Vancouver's raining frogs and alligators spring days. I boldly wore my fine Italian leather Mary Jane wedge heels, dark skirt and blouse to the interview underneath my beige trench coat. I left at least two hours early to take three buses to get to New Westminster. I was a mess when I arrived. My feet were soaked, the leather shoes water marked. My eyeglasses were fogged up. My hands were cold and clammy. The cotton trench coat I wore was before 3M made more water repelling fabrics.

Mr. Smith worked out of a low-rise brick and mortar building with fluorescent light fixtures and utilitarian furnishings. My nerves were fried. I entered, quickly excusing myself to the washroom. Without a comb in my purse, I ran my hand through my long hair, and wiped the water off my face and eyeglasses with toilet paper. I regained my composure and was ushered to sit on a chair until Mr. Smith was available. The clock turned 3:00 p.m., and I was still sitting. Mr. Smith's secretary was not around. I waited another half an hour. Then the secretary came to the reception area and informed me that Mr. Smith would now see me.

He was a tall imposing grey-haired man in his mid-50s, who hadn't been to the gym in some time. Mr. Smith wore a white shirt with a dark conservative tie and dress pants. He sat at his desk with a slight scowl on his face. I walked over to extend my hand to shake his. He declined. There was another engineer in his office. The younger engineer with a full head of blond hair sat in a chair besides Mr. Smith's desk. He was about to get up from his chair to shake my hand, but hesitated for a moment, and sat down.

Within seconds, I knew Mr. Smith was livid with me. His first question was "How much do you weigh?"

I lied. "One hundred pounds." I was more like 96 pounds soaking wet, if that.

Mr. Smith retaliated, "Well, don't you think you're rather small to be an engineer?"

He leaned back in his chair, and put his feet on his desk. I was flabbergasted and sat there, my mouth was on mute. Mr. Smith continued his rant, "Listen, we can't hire you. We can't send a girl to the field."

I sat there, defeated.

All I could hear from Mr. Smith's mouth were these words, "Those men swear a lot. How are you going to deal with this?"

I was a meek, gutless wonder. Any fire in my belly had been extinguished on the bus ride over here. I flew into post-traumatic stress disorder. I froze. He continued his rant. I did the unthinkable. My eyes swelled up. My tears started flowing, at the most inopportune moment. Then I started sniffling. Damn it! This is why men don't want women in engineering. But Mr. Smith continued with his female bashing. His ego was crushed. He wanted to make it clear that I was a piece of shit and had no right to waste his time further. I don't know why I didn't speak to him like the guys in my engineering class spoke, with the f-ck word as a noun, subject, predicate, or adverb.

The younger engineer stood up for me. He made a feeble attempt to get me hired. He reminded Mr. Smith that I could help out with the drafting, which was in those days hand drawn on isometric vellum paper using F-leaded pencils – using plastic triangles, T-shaped rulers, and flexi-curves. After wards, the drawings would be outlined in ink by special drafting pens.

Mr. Smith had already made his mind up the moment he saw me enter the room. He retorted, "No."

I had reached the point of no return. Mr. Smith ordered his younger engineer to show me the way out. As I left, the younger engineer, apologetically said, "I'm sorry."

Afterwards, I commiserated my interview experience with another female engineering student. I did not get the empathy I would have expected. Her words cut through my fragile psyche, "How could you be so stupid! Girls can't cry. That's why guys don't want girls to be engineers!"

That was not the sympathy I was hoping to hear. Her words only showed how much stress all we girls felt being in that first batch – struggling to be perfect when we were only human.

Smith Ramsey Kennedy was the only summer job interview off campus I got. I couldn't take further rejection in industry, so I looked for jobs on campus. It would be easier and less stress. Wandering absentmindedly around campus, I ran into Gary, my second-year electrical engineering course lab partner. Little did I know, he had a crush on me. Gary finally asked me for my phone number. My mother wasn't thrilled about me dating Gary. He had outrageously long hair and dressed in jeans and t-shirts, which didn't impress her. His roommate Bill wasn't seeing anyone. Suzanne, my Hungarian girlfriend who was 22 going on 30, went after him.

Chapter 18 –
Tale of a Bollywood Bride – Pritha

Living on my own gave me a profound independence and a sense of well-being. By third year, most guys in my class were living on their own. My peaceful existence was interrupted at 2:00 a.m. with desperate banging on my front door. It was Pritha, a striking East Indian girl from my art class. I had not seen her in the library, since last year. Shivering in bare feet wearing a hospital gown, she needed ten dollars to pay for a taxi ride. I complied. Her parents found out that she was seeing this other East Indian boy. The problem he was from a caste higher than the one her family came from. When they found this out, Pritha's parents put her under house arrest. Her parents were going to do what's right for their daughter and marry her off to a nice East Indian boy from the proper village.

Desperate people do desperate things. Pritha grabbed a pair of scissors and cut her tongue. When her parents saw her blood oozing from her mouth, they assumed a worst-case scenario and that she was deathly ill. They rushed her into the emergency room at the Vancouver General Hospital. The doctor on call knew right away what was going on. The bleeding was minor. He told her that she would have to check out in the morning, as there was nothing wrong with her. With no other place to go to, she fled to my place. We talked for the next hour. She had no money to her name. All the money she earned working part-time at the grocery store was seized by her parents. She had dropped out of university. At 21, she should have been married off by now. It was contentious enough that she had the gall of going to university. Education was not seen as a pre-marital requirement.

There we were. I asked her, even if she crashed at my place, then what? The cops would be looking for her. She did the unthinkable. She called up her parents. They arrived within the hour, dressed in traditional Sikh clothing. They barged through the front door in procession to remove Pritha, as if she was some sort of chattel, from my apartment. Not a word was spoken to me. Their eyes sneered at me, as I was seen as a co-conspirator in their eldest daughter's life. That made me feel like I was the bad guy. I never heard from Pritha again. She never paid me back the money she borrowed. I knew that.

I only mention Pritha in my memoir because her goals in life were diametrically opposed to mine. As a friend, all I wanted to see for her was that she be happy. Likewise, she wanted me to be happy and from her perspective, I needed a man. From what she could tell what was stopping me was not having a big rack on my chest, like hers. She nagged at me tirelessly that I invest in a bust developer, which were the rage back then. For $3.99 plus shipping and handling, I purchased this blue plastic contraption with two hand grips separated by a spring-loaded bar. The device arrived discreetly in an unmarked plain brown paper box. For 30 days, I pressed on that thing 30 times a daily, gullible to think that something miraculous would happen. It's virtually impossible to add something to nothing. After 30 days, I got nothing. Not all was lost, my more endowed younger sister bought the contraption from me. Then and there, I knew I was already becoming an environmentalist, holding a high moral ground on re-cycling.

When it came to fourth-year engineering, our small family of bio-resources engineering students was just that – a family of five engineers in our year. We all sat in an upstairs room in the second floor of an engineering annex. While all the guys in my fourth-year class had dates for the Engineers' Ball, I had become so accustomed of doing things on my own that I just went by myself. We were a close group, but after graduation, our paths would not pass at all. Coming from such a small faculty, our presence was a speck in the scheme of things. Eventually, the department of bio-resources engineering got scooped up by the chemical engineering department.

I finished my fourth year a term early, as I received some transfer credits from BCIT. Since I was no longer a full-time university student, I had to quit my lucrative part-time job at the university library (which my big sister had helped me get). I did the unthinkable for the next four months. I was emotionally exhausted from studying engineering. I didn't want to start a revolution. I just wanted to work as an engineer. What I did next was procrastinate.

It was on my last shift working for the UBC Library, that I ran into Pritha's younger sister. She told me that Pritha went back to India to find a nice boy to marry to no avail. Finally, their father realized how unhappy his daughter was and Pritha married her boyfriend two castes ahead of her. I'm assuming she lived happily ever after, because every great memoir needs a fairy tale (spoiler alert, you can't quit reading and ask for a refund now because mine isn't ...)

Chapter 19 –
Pink Collar Ghetto Blues

Engineering school came to an end. I attended convocation 1978. Out of the 14 women who entered engineering school in 1974, five of us graduated. I wasn't ecstatic like the guys in my class, so proud on entering the profession as it would get them good jobs, but more importantly, help get them laid. I wasn't even excited enough to bother attending the iron ring ceremony in 1978, along with my engineering class – a once secretive and closed ceremony held in a dimly lit lower room of a lowbrow hotel.

Conceived in 1922, the iron ring is presented to Canadian engineering graduates to symbolize the pride that engineers have to uphold and their obligation to live by a high standard of professional conduct. The Canadian iron ring was once made of iron, but in recent years, fabricated out of stainless steel. Such a ring worn on the pinky finger of a working engineer has a multi-faceted exterior surface, so that the ring can drag against the surface of the drawings, the engineer may be viewing. In 1970, a similar iron ring tradition was established amongst American engineers under the Order of Engineer, except such stainless steel rings have smooth exteriors.

There I was in the spring of 1978. I had run out of gas. I had no road map. No direction. No mentors. My only connection to a female engineer was a measly three paragraph blurb in a brochure I found in a university counsellor's office about this tall unmarried Russian woman with oversized black glasses and a false sense of fashion who worked for B.C. Hydro. I was on pause. I couldn't delude myself. I had bills to pay, with no sugar daddy in sight. I needed to work. But I sure wasn't going back to working for Peter's or in retail.

Things could have been way worse. At least, I could keep my clothes on. I signed up with Kelly Manpower, an office temp agency for work. I still remembered some of my Pitman shorthand from high school. My typing was a pass. My filing was 100 per cent because they had penciled in numbers on the corners of the filing cards I was tested on. I only worked for a few weeks in the pink collar ghetto, before I bailed out. I would deny being a feminist, back then and even now. I wonder how happy one can be, depending on another person, like a husband, to take care of you. Or going through the motions of working at a job you hate, until someone rescues you from some misfortunate circumstances. Shoot me now or whenever.

I survived two office temporary assignments. The first was stuffing envelopes (ccoincidentally, for the B.C. Provincial Engineering Association) in a basement room with three other women – all high school drop outs, who didn't make the semi-finals in the beauty contest. Single, over 30, their glory man-hunting days were over. Waiting in the reception area, I squirmed to hear a thin mousey woman in her mid-20s share with a girlfriend on the phone details of the clandestine relationship she had with her boss, an older married man.

I must not end up like my mother, I told myself. However, it would be my mother's fantasy for me to end up like Bridget and Simone, two high school classmates who resided in the Shaughnessy district. Both lived in houses like the one I had scrubbed weekly years before. When Bridget's mother came by to pick her daughter up in her white

Mustang convertible, my mother jumped out of the house to meet her on the street. My mother savoured every word that Bridget's mother had to say. After all, Bridget's father was a doctor.

The most popular girl in high school, the blonde blue-eyed princess was the dentist's daughter. Like her mother, she would follow suit, also marrying a dentist, too. Like her mother, Caroline would also have a blonde blue-eyed princess with long straight blonde hair and 40 years later, end up living the country club life in a southern American state. She was well preserved for her age – despite the blank look between her eyes as she described the good life she led at a high school reunion.

My last assignment for the Kelly Temp Agency was working as a Girl Friday for a small engineering company. When companies were too cheap to hire a full-time secretary, they would let the work pile up from Monday to Thursday. And I, a Girl Friday, would come in on Fridays and work on a multitude of pink collar ghetto tasks. Being the Girl Friday had no prestige or status in the office ranks. Typing was not my forte. Banging on the IBM Selectric typewriter, it would take me an hour to type up a letter. I worked there for two consecutive Fridays and was never called back a third time. That suited me fine. I was told that they realized they needed a full-time person. I am pretty sure I was fired. No hard feelings, honestly.

It occurred to me that I had accrued some student loans. With burger flipping jobs, paying my loans off would take until perpetuity. While most of my peers already found jobs before they graduated, I had not. I never entertained any of the pre-graduation job opportunities very seriously, as I didn't realize that's what you're supposed to be doing. I was chasing a train that already left town. While my peers went to work immediately after graduation or were taking the summer off to travel Europe, I took a contract position with the department of civil engineering, working for Dr. W.K. Oldham. I saw the posting in the hallway. Nobody else applied. All the guys found better jobs elsewhere in industry. Dr. Oldham's actions spoke louder than his words. Over the years, he quietly stuck his neck out for female engineering students.

When the contract ended, I was living in my upstairs apartment baking whipped cream-filled profiteroles and painting to my heart's content. I had entered my art in some juried competitions and was accepted. I sold a couple of pieces of canvas, but I was quickly running out of options. Life was good, but I knew sooner or later, that I needed to find a real engineering job.

The serious engineering job interviews I received were jobs that nobody else in my class would be interested in, like working for a Mexican non-government organization or working in Bellingham for a major wastewater treatment engineering company that would eventually grow up to be a kick ass multi-national engineering conglomerate of repute. The American engineering company already had a woman engineer. She worked 60 to 80 hours plus weekly to impress her boss and prepare herself for state engineering exams. I had enough of the nun's existence, eat, sleep and study. I declined the offer from CH2M Hill. Maybe I was a fool. I would have eventually ended up in Canada, likely Calgary, as CH2M Hill set up shop there in a few years. My big sister snarled at me for even considering work in the United States, and this was decades before Donald Trump was elected.

PART TWO –
WHO THE HELL DO YOU THINK YOU ARE?

Chapter 20 –
Lotus Land to Cowtown

At 21 years old, I was mesmerized by the soothing voice of Franz Harpain, a chillaxed math teacher at Langara College, who hooked me into the idea of learning transcendental meditation. TM was supposed to solve all of the world problems, but first, would do wonders for me. It cost ten dollars to be initiated into TM. Since I could never really get good at drinking like my engineering peers, I spent my spare time at social gatherings with TM freaks. It was at one of these meetings, I met Laara Zatorski, an effervescent blonde Ukrainian divorcee who lived and worked in Calgary. We exchanged phone numbers, and she invited me to crash at her place if I were to come to Calgary.

I took up on her offer. There was a ninety-nine-dollar flight seat sale, and I flew to Calgary for the weekend. Laara shared a duplex with two other women, all gainfully employed. Laara was a larger than life secretary. She was this extroverted big sister I never had, who had been around the block, several more times than I. That weekend, Laara and I went to a house party hosted by one of her accountant friends, who really loved his liquor. While she hollered at everyone like a hyena, I was a wallflower. Then my eyes locked on to the eyes of a shorter version of Clark Kent, a young man who was moving from Winnipeg to article for a Calgary downtown law firm.

My heart fluttered. He gave me his business card and told me to call him, should I come back to Calgary.

I picked the wild card. Pauline, the tea cup reader who worked out of the Sportsman Café in downtown Vancouver, had predicted months earlier that I would be surrounded by boxes. She was right. There was nothing for me to stay in Vancouver. I didn't get along with my sisters. My childhood memories were a dark cloud raining over me. I had been treated with allergy shots for five years and that program wasn't working. The only thing left Dr. Mandell could prescribe to rid myself of hay fever is a change of residence to get away from Vancouver alder and birch trees. Pauline predicted that my residence would either be in another province or in the United States of America. I packed up all my belongings into a trunk, my bicycle and twenty boxes filled with engineering books, and boarded a train to Calgary.

I was scared and excited at the same time. I had no plan. Quite frankly, I was on the lam from Vancouver, in the same way my Uncle Victor left for Calgary decades earlier. Laara convinced me that Alberta was the land of milk and honey, and the roads were paved with black gold. I rationalized to myself that with an unemployment rate of four per cent in Calgary vs. seven per cent in Vancouver, the odds of finding work would be better in Calgary than in Vancouver.

The train rolled into Calgary when the sun had gone down. Jim Stephens, the lawyer who I met two months before told me to call him. He was congenial but the spark he had initially shown me had been flushed down the toilet. On the drive to Laara's house in his big American car, I learned that there was some significant other he failed to mentioned – a girlfriend, decided to move from Winnipeg with him after all, was now living with him in Calgary. But he was a man of his word, as he helped unload my luggage and welcome me to Calgary.

I stayed at Laara's house until I found myself a place to rent – a furnished one bedroom upstairs suite in the Beltline, an inner city Calgary neighbourhood. Yes, its décor was trailer park trash with the plastic laminate walls, mustard shag carpet from the decade before, a warped saggy bed, a fridge with the ice box filled with so much ice that the door couldn't close and a crotchety gas stove and galley laminate counter. The rent was affordable, while my neighbours were questionable. Drunks would pound on the main door on Saturday evenings. I didn't need to buy weed. Pockets of second hand marijuana smoke in the air was a regular occurrence in this inner city neighbourhood. My landlady, Madame Hebert, was an 87-year-old, who partied every Saturday night with the local Alliance Française, until 2:00 a.m. Where's an unemployed girl like me supposed to crash? At least, I wasn't living in Bowness and Forest Lawn, considered by elite to be uninhabitable for them.

I had arrived in a town that year, like thousands of other college graduates, from across Canada, in particular, Ontario and Quebec, who Mayor Ralph Klein later fondly referred to as "Eastern creeps and bums". My trunk with the brass ornaments became my coffee table for wherever I moved to. My red and white Raleigh bicycle became my mode of urban transportation, even for evening soirees, dressed in a silk and linen pantsuit. My summer evenings were spent sprawled over a bath towel on the grass in the elementary school across the street, reading newspapers and books. I was quenched for knowledge and was a big fan of self-help books. As by now, you must have figured I was a therapist's dream come true.

My Uncle Victor, an entrepreneur who had an office on the second floor of a Calgary downtown hotel was kind of enough to let me borrow an electric typewriter to type out my resume; which I would get photocopied on high quality cream-coloured linen-textured paper. A formally dressed businessman, my uncle wore white shirts, dark custom-made suits, polished shoes and drove a Cadillac or

Lincoln. He walked around like he owned the world. Shove a cigar in one hand and a glass of Cognac in the other; there you have it – someone who can be construed as a genius or narcissist. (He was old school, but it would be decades later, I would learn he always had his heart in the right place.) On his right hand, he wore an oval ring, inlaid with the jade from the ring my grandmother wore.

Uncle Victor laughed at me, "Woman engineer. Who's going to hire you?"

I continued typing, head down, under my breath, "Someone."

As I talked back to him, he remarked, "You talk like that, you're going to get fired."

I swear my uncle was so intuitive, that you would call him psychic. He read people like an open book. Coming from the bowels of Chinatown, he was ambitious, an alcoholic, and a workaholic. Like everyone from the family I came from, work was synonymous with play. There was no divisible line.

After mailing out three hundred resumes and getting diddly squat, I literally pounded the pavement. This was not as horrifying a task to fulfill as I thought. As a teen, I once went door to door around one city block selling Regal Greeting cards from a catalogue. In the hallways of Calgary office towers, I was totally inspired by the unkempt construction dude with the missing front teeth, who wouldn't flinch when potential employers would tell him straight up they couldn't hire him. Undaunted, he walked down the hall and knocked on another door.

I had a list of doors to knock on. I dressed for success in my *Suzy Creamcheese* blue and white narrow pinstriped skirt suit and white cotton tailored blouse while teetering in my Italian brown leather heels (which I brought on sale at Ingledew's, an upscale Vancouver shoe retailer, which closed down after 100 years in business in 2017). The suit was so well tailored in Montreal, when women's clothing was still made in Canada, that in the years to come, I received more compliments from the men than the ladies. I carried my resume in an Italian burgundy briefcase with shiny gold-plated combination locks – a graduation gift from my big sister that cost her two hundred bucks.

Calgary was booming. I got interviewed on the spot for a civil engineering firm. The hiring manager was a stereotypical middle-aged receding hairline Caucasian male wearing the oversized 1970s eyeglasses, white shirt with a striped tie, and creased grey flannel slacks. I figured I had a shot with him. Well, that's what they all say about newly-minted engineering grads – a bit cocky. I was an engineering graduate with two years of experience including my one year for Environment Canada, but my strength I thought was my summer job at Kerr Priestman. Mr. What's His Face was bragging to me about his recent acquisition – a University of Waterloo engineering grad who stood six feet tall and had a black belt in karate. I couldn't argue about the University of Waterloo grad's credential – as that university was the most sought-after in the nation, akin to Canada's MIT. I lost this beauty contest in terms of height. As far as the talent contest went, karate was certainly not part of my repertoire.

But I did have something, that perhaps, this employer wasn't prepared for when I responded to him, "Engineering begins with the shoulders up, not the shoulders down."

These exact words were ingrained in my brain by Tony, a Canada Manpower Counsellor who always had time to encourage an underdog like me. The hiring manager turned a bright beet red. I stood up, excused myself and wished him a nice day.

I could never be prepared, so I spent one Saturday morning at the Alberta Career Centre, which offered workshops on resume writing and job hunting. Coincidentally, it turned out all the participants were women. Beverly, the outgoing middle-aged woman who led the discussion was as open and frank as the questions that came her way.

One of the women in the room asked, "What do you do when they ask about your sex life?"

Beverly replied, "How badly do you want the job?"

The woman nodded, "Kind of badly."

Beverly proceeded, "Well, I was in an interview. And I knew the guy was a jerk. But I really needed the money. He asked me how often I had sex. He was worried that I was going to get pregnant and run off on him."

We were all ears, as the other woman asked Beverly, "What did you tell him?"

Beverly answered, "I have sex once a week."

Our eyebrows were raised.

Beverly nodded, "Yes, Sir. Just on Sunday mornings. So I can go to bed early on Sundays and be ready for work bright and early on Mondays."

One of us asked her, "Did he buy into your answer?"

Beverly shook her head, "No. He said to me, 'Really, only once a week?'"

We gasped.

Beverly added, "I lied to him when I told him, 'Yes, only once a week.'"

I asked, "Did you get the job?"

Beverly replied, "Yes, I did…I knew he was going to be a real ass. I kept job hunting and found myself a better job three months later, and quit."

A woman engineer, who scored a job with a major oil company swore by two books that were gospel for our survival. John Molloy's *Dress for Success for Women*, advised that women dress like flight attendants or persons of authority – matching knee length dark suits in black, navy, grey or burgundy to match corporate annual reports with plain white cotton or silk blouses/shirts preferably with buttoned down cuffs and pointed collars. High quality pumps were to be plain with a short heel and Italian leather. I adhered to John's book to a 'T'. I learned that it was advantageous, making it easy to dress for work – three suits and five blouses did the trick.

The next book of interest was Betty Lehan Harragan's *Games Mother Never Taught You*. It was nowhere to be found on the bookshelf on the Coles Bookstore in Calgary's downtown TD Square. The only way I could buy a copy was by asking the clerk. She nodded her head and deftly unlocked a special drawer and pulled out a copy for me to buy. Coming from a lower working class family, these two books replaced the big brother, father or uncle from a professional or upper class family to guide and mentor me. Whether Betty was right or wrong, her insights gave me the education that I didn't get from university. Of course, she could have placed more emphasis on pedigree, as years later, I shall see.

Chapter 21 –
Cube Farm Sewer Rat

Within 90 days of arriving in Calgary, I was hired by Reid Crowther & Partners, which would later become part of a multinational engineering company called Earthtek. I accepted the offer, which was ten per cent less than the going rate. I had arrived from BC and didn't know that Calgary engineering grads were paid more. I was hired by Mark Davies, a British born and trained engineer, who wore a three-piece vested suit, and resided in the corner office – a piece of real estate that epitomized success for baby boomers. My heart sunk when I was shown my workspace – a six by four square foot cubicle – no window along with all the non-Brits – East Indian, Chinese, Polish, Hungarian engineers and draftspersons. The exterior window offices were occupied by tall white males imported from the United Kingdom.

My first job involved sizing sewer lines for new residential neighborhoods and sizing sewage treatment lagoons for a project in Cuba. After a year, if I lasted, I would end like Sudeep, plotting pump curves and sizing pumps. Well, suck it up princess, I told myself. All the non-Brits were dissatisfied with their lot in life. Sudeep, new to Canada from India, had his MBA and over 15 years' work experience. Pat Lee, raised and educated in Manchester, England, had the right accent but the wrong ethnicity. He disliked Calgary's bloody cold winters and longed to move to Vancouver and open up a res-taurant. Terry Jaworski, the long-haired Polish draftsperson, didn't mind the work but wanted to make more money.

The year was 1979. Donna, our secretary with four grown up chil-dren, doted over the men like she was their mother. Precisely, at 10:00 a.m. and 3:00 p.m., she would enter our work area with a tray of coffee and tea. She would ring a little brass bell. Work ceased for the next 15 minutes. She had her order down pat. Some men liked their coffee black, others with two lumps of sugar, no cream and others fully loaded. One English man preferred Red Rose pekoe tea. I declined patronizing her. I thought her job was pathetic. I just drank plain water instead. The next 15 minutes were painfully long. Nobody in our marginalized group would volunteer to speak, for fear of repercussions. We faked our smiles and artificially laughed at Mr. Davies' corny jokes.

There were two Ukrainian engineers – Wes Marchuk and Tony Hnatiuk, who worked in the construction department – the only ones who could get away coming to work in blue jeans, every day. My roommate at that time, had her signature dish, which she was to prepare and that I could consume on a regular basis – borsch. Wes and Tony who were assigned to adjacent cubicles, took glee in their findings and would gladly purport, "Oh, Nattalia, you need to step inside into our office for a chest inspection."

I smiled to myself at first, hoping that their words of wisdom would be a one-time thing. But I was wrong. Their request for a physical examination would become a constant. We would giggle at one another, as I knew they were teasing me. Wes and Tony got progressively friendlier, too; presenting their argument for a chest inspection, "You know, Nattalia. Borsch makes you grow hairs on your chest."

Then one day, our paths traversed the open concept drafting department area which seated at least half of the company's employees. Tony blasted at me, "Hey, Nattalia, do you have any hairs on your chest? We need to do that inspection."

I had no what idea that overcame me. It was like spontaneous combustion. I hollered back at Tony, as he passed me in the hallway, "No, Tony. Men grow hairs on their chest. Women grow boobs."

The whole office was in stitches. From that point on, neither Tony nor Wes, requested a chest inspection from me.

When I started at Reid Crowther & Partners, I assumed that my career with them would be until death or retirement do we part. Assigned to the sewage and wastewater treatment group, I literally had a shitty job – sizing sewer lines for expanding Calgary subdivisions. I reported to Peter Kennedy, a tall prematurely greying Brit, fresh off the plane from the United Kingdom with a very pregnant wife.

The first week on the job was tolerable. After that, it was outright boring. As an engineer-in-training, my job was to do the work the senior engineers turned their noses up at. The City of Calgary was the bread and butter client for Reid Crowther & Partners. Everybody in our group was working on some aspect of this municipality's sewage or wastewater disposal system. Sudeep and Pat spent their days calculating water heads to size pumps. Both Sudeep and Pat would escape their mundane existence by going for smoke breaks. For any of us, a promotion would be to work in the water treatment group, which sounded a whole lot sexier.

I was lucky. I was doing real engineering work. The black-haired Russian woman who claimed to have an engineering degree worked alongside the Hungarian draftsperson. Anna, a dark-haired Italian girl with a university degree in chemistry, was filing papers.

The engineers before me complained about being underpaid, mainly because they were immigrant engineers and didn't have enough Canadian work experience to qualify. Except for Sudeep, the darkest skin of them all. He had his Canadian Professional Engineering status, but just kept everything to himself and just went along with the condescending attitude management had towards him. Pat didn't think he had anything to lose about voicing his opinion that he was underpaid. Mr. Davies took matters in his own hand. He returned the following Monday at 3:00 p.m. and plunked down on Terry's drafting table a box of zucchinis from his garden. Mr. Davies uttered some words feebly under his breath that the company was not in a position to give out raises.

One by one, the zucchinis were unpacked and laid out along Terry's drafting table, then measured – anywhere from four inches to almost 12 inches. The way the zucchinis were literally man handled, one knew exactly what references were made.

The worst part about my job was there no reprieve from being trapped in the cubicle farm. There was no field work. I was Reid Crowther & Partners' first woman engineer the company ever hired. Everybody was shocked, but I knew Mr. Davies was desperate, as Calgary was booming. A year later, I was fired. Hans Krueger, a tall dark handsome (but then, we were all good looking because we were young) University of Waterloo grad, strolled into Reid Crowther & Partners' office. Mark hired him in a heartbeat. I was told to show him what I did, as a promotion would be coming my way, or so I thought. Instead, Peter, who was now a father, called me into his office to inform me that I was being let go.

I asked Peter for a job reference. His exact words were, "Nattalia, I don't believe women should be engineers. I think women should be nurses or teachers, like my wife Maria. I will gladly give you a job reference, if you agree to go back to school and become a nurse or teacher."

Betrayed, I conferred with Sudeep. I was worried that now I had no one to vouch for the engineering-in-training work that I did. He reassured me, "Don't worry. I'll sign for you."

I had come this far. Damned, you bet I wanted to get my Professional Engineering status. Hans was devastated about my job termination, when I was given a working notice -- no severance, just 30 days to stay in the office and work, like nothing happened.

I didn't leave Reid Crowther & Partners crushed. On my last day, I walked into Mr. Davies' office, and gave him the Spanish Inquisition. My ego wasn't going to take this blow lightly. I asked him why he hired me, if was going to fire me a year later. He trembled in his seat and was tongue tied, until he told me, "I was really impressed by your qualifications."

I asked him, "Why are you terminating me?"

Mr. Davies really stammered and didn't have a precise answer. I left head high and ego intact.

Trouble happens in threes. My beloved red Fiat, (a.k.a. Fix It Again Tony) was decimated in a hit and run accident, and had no future except a scrap yard. Roger, the tall dark handsome Texan lawyer, and son of an oil baron, broke up with me and ran off to California with his ex-roommate's ex-girlfriend. My roommate, a nurse from Winnipeg, kicked me out because she was worried that I wasn't going to pay my share of the rent.

Working as an engineer was all I obsessed about. Some non-engineering friends thought that I had a classic case of gender discrimination and suggested it was time that I take on the role of a social justice warrior and spend what money I had on lawyers. The best advice I got was from George Atkins, an older engineer who had a real job working for an oil company. He took pride in going to work with a beaded choker around his neck and hair longer than his collar, "If you do that, you will never work as an engineer again."

He was right. The engineering community is as tight as Crazy Glue. I went to a job interview for a mid-sized oil and gas engineering company. The interviewer asked me if I knew Jane Doe, another woman engineer, who I heard slept around. I was flabbergasted. Here was an engineer who went to university in eastern Canada and worked in Calgary. He knew about Jane who worked and lived in B.C. How could this be so? I feigned ignorance in the interview and answered all of his questions to the best of my ability.

However, I realized by the end of the interview, I was told little about the job position. I asked the interviewing engineer, "Can you tell me more about the job?"

He replied, "What job? I'm here to ask you out on a date."

This had to be one of the low points in my life, or so I thought. My pride was broken. I was ashamed and afraid of what others were thinking. There were scores of newly-minted female engineering grads, married to equally successful male counterparts, living the Calgary good life, working for oil companies earning the big bucks and being showered with bonuses, complimentary tickets to hockey games, company expense accounts and trips on the corporate jet. These women were still working. I had met a handful of them over dinner once. I was the only one who had been involuntarily terminated. What was wrong with me? Doubt Ville took over the grey matter in my head. Worst yet, I was two years older than everyone and still unmarried.

The new place I rented had an automatic dishwasher, but that appliance had a price to pay. My rent was double previously. I hid in my apartment for a month, waking up each morning, bawling my eyes out. Wes Marchuk, the divorced construction engineer in his late 30s from Reid Crowther & Partners, counselled me. I felt like I had reached rock bottom. I was ashamed to tell my mother that her bigshot daughter was now on unemployment insurance. The picture I had painted in my mind wasn't as black as it really was. I was still young, cheap to hire, and companies were looking.

Fluor Daniel, a major international Irvine California-based engineering, procurement and construction company, was setting up shop in Calgary in anticipation of being awarded several multi-million projects. They had posted an advertisement that they were holding a job fair in Calgary at a local hotel one weekend. I walked in with my resume. We sat down. They interviewed me on the spot. I had my fill as a wastewater treatment and sewer design engineer.

Chapter 22 –
Company Man, Tiger Lady

Bob Bruce, the Fluor hiring manager, was a big American engineer. He stood at least six feet tall, carried a few pounds too many and had been with the firm since he graduated. He worked on over 20 projects and over six continents. He never stayed in one place for more than three years and subsequently, changed wives as often as he changed continental workplaces. He was a company man, who wore a Fluor silver pin with three diamonds and one emerald, for his dedication. A weird question Bob asked me as I got up and walked away, "Are you a ballerina?"

That was meant to be a compliment, although I was somewhat perplexed. To my surprise, I got called back to Fluor's office for a second interview, which proved to me I had entered friendly territories. It included a lunch at Chi-Chi's, a franchised Mexican restaurant, because all the senior people were from Houston and liked Mexican food. I was being primed for a scheduling engineer position. I reported to Carl Jones, a bald portly engineer from Houston, who was sent up to Calgary to convert us into the ways of Fluor Daniel. His effervescence bubbled over daily with his southern drawl, "How are y'all doing?"

Carl had been with Fluor over 20 years. He was a company man and loving it. He was one of the luckier ones, who will likely stay with Fluor until death or retirement do he depart. He was happily married to his faithful high school sweetheart, who was a stay-at-home wife and took care of their only child, a son.

I reported to Dave Cairns, an engineer who had left engineering temporarily, to do something that was lucrative enough for him to buy a Mercedes Benz. Dave and I clicked. He was a truly nice guy. Guess what? Fluor paid much better than Reid Crowther & Partners. I was hired along with two other new engineering grads – Adam Shore, a tall handsome kid from rural Alberta and George Brown, an engineering grad from the Maritimes who married his high school sweetheart, a redhead with wavy locks named Fiona. With his engaging personality, tall bod and good looks, Adam, forty years later, would excel in his career. But of the three of us who started out at the same time, only George would become a lifer at Fluor and walk around the office with a Fluor pin with three diamonds and one emerald.

I may have been one of the first women engineers to be hired at Fluor in Calgary. I wasn't alone. There was a super smart cost engineer who graduated from a Montreal university and was very active networking at the Calgary Chapter of the Project Management Institute. Within weeks of me being hired, a freshly-minted University of Toronto female engineering grad was hired. Then there was another female engineer with her MBA who worked in the piping department.

Working at Fluor was refreshing compared to the stiff upper lip British management style of my previous employer. I worked with engineers from around the world. Fluor was an international company that had worked on projects nearly on every continent, and its workers reflected this diversity. While male engineers in management received a clothing allowance for a new suit every year,

I took the plunge and forked out the $600 for an Al Murano tailor-made navy suit. It was worth every penny. I felt like a super hero at work. In public, I looked like an Air Canada flight attendant.

Chapter 23 –
Monday Day Fever

Easy come, easy go. Gossip had it that Roy French, the troubleshooting scheduling engineer from California, volunteered to come work in Calgary to get away from one or perhaps, all of his ex-wives. Tall, slender and handsome with pearly white teeth, and long wavy blonde hair, he looked like he came offstage singing besides the Bee Gees. Everybody anticipated that Fluor was going to get the big job for the Cold Lake Expansion Project – 10 frigging billion dollars – that was the largest amount of money to be spent on any project. The black gold rush was in swing. Roy didn't come alone. He brought his entourage – first, with his right-hand man – a tall moustache beer-bellied engineer who just went with the flow. Some say he just came to work stoned all the time. I believe them, as his mind seemed to be elsewhere. There was also, Brian Church, a very easy on the eyes single engineer, who looked like he stepped off the cover of a men's fashion magazine.

When Brian arrived the first Monday of a new month, he didn't know that the girls in the secretarial pool were fighting to get into his pants. The prettiest, fairest and blondest, was Lisa. She was this hazel-eyed cutie pie, who staked her claim, the moment Brian showed up, "Stay away from Brian, you bitches. He's mine."

She wasn't kidding. First, Lisa started loitering around Brian's desk, making sure she was close enough for him to catch a whiff of her sexy cologne. (I wonder if she wore N° 5 CHANEL?) She gave him prolonged eye contact. I witnessed her deliberately leaning forward as she dropped some papers on his desk, so some of her assets could be sized up. It wasn't before too long that he asked her out. She did well. I heard through the grapevine that he did marry her.

The temperature in the office was rising with the increasing number of new hires and potential after hour nocturnal activity amongst those attracted to each other, regardless of marital status. Dave nicknamed Fluor *Peyton Place*, after a primetime soap opera, of that era. Coffee station talk gushed with office affairs and romance, while Fluor upper management was more concerned about employee retention.

As fast as Fluor was hiring, people were leaving for more money. Every few months, upper management surprised us with pizza. They must have seen us as lab rats and that pizza would keep the indecisive ones in the cube farm. I guess they knew that they didn't have to worry about me.

My two contemporaries, Adam Shore and George Brown, sat in adjacent cubicles to mine. I was older than them. I had more engineering experience than them, but that didn't count. When we all started, we earned the same salary. With the looming staff shortages, management offered us raises. Both Adam and George got $300 a month increase, I got $50. I was disappointed and confided with Dave. He told me that this was not acceptable and ordered me to discuss this pay discrepancy with Peter Droeker, a sharp South African engineer, three ranks my superior. Dave got me all wound up, without even drinking coffee. I didn't hesitate for a moment. I put on my custom-made jacket to my navy suit, tied my hair back in my pony tail and tightened the scarf around my neck. I stomped over to Peter's office, who maintained an open door policy. I stood at the front of his door. He motioned me to enter, "Yes, Nattalia, come in."

I plunked myself down and cut to the chase, "Peter, I notice that George and Adam got $300 per month raise and I got only $50."

Peter flustered, blurted back at me, "Well, we can't really give you more money. Fluor is struggling. We've got to watch our cash flow."

I smelled bullshit, in the same way, we know bears poop in the woods, too. What happened next was unthinkable. Perhaps, this was an act of God, because I am still flabbergasted. I nodded to Peter and walked to the door, paused and turned around to Peter. I was still standing and I taunted him. "Peter, I know where I stand."

I went back to report to Dave. He was disappointed, too. Dave suggested that I go find another job, which I did not want to do because I was desperately trying to establish myself. Two weeks later, I got a call from Bob Bruce, who was four ranks my senior, "Nattalia, we're giving you that $300 per month raise."

I asked, "When?"

He replied, "Next pay period." Now, I could have asked for it be made retroactive, but I was just glad to get paid as much as George and Adam.

Peter told me that one of the procurement managers complained that I was always on his back to meet his deliverables and that I was being called a "tiger lady." I asked him if I should back off. Peter said, "No way ... you're doing a good job."

During my tenure at Fluor, I worked on the Union Carbide Project, the Ultramar Refinery Project and a methanol plant in Medicine Hat. We only got one field trip. It was a day trip to Medicine Hat to visit the methanol gas plant. It was a day to remember. We were so high above the ground. We climbed up the grated staircases to the top of the cooling towers. You might say our heads were in the clouds.

Behind my back, one of my informants, Dianne, an underemployed geologist from the Agency told me that she had a heart to heart talk with Roy French. Even though I shone in the boardroom with streamlining a change order work flow process, my efforts were not rewarded like my male peers. Roy told Dianne, who told me, that Roy told her, "It's too bad Nattalia is a woman. She's not going to go anywhere at Fluor."

He wasn't kidding. Both Adam and George were sent to Irvine for additional training. I wasn't even considered.

Chapter 24 –
First Bust

Everything was going great at Fluor, until Prime Minister Pierre Elliott Trudeau announced the National Energy Program in 1980. There was this imbalance in national wealth across Canada and Trudeau thought it was time to play Robin Hood. It was bad enough that unemployed and under-employed persons were flocking to Alberta for work, let alone bleed Ontario and Quebec of taxpayers they needed. The National Energy Program sent a cardiac arrest through the oil industry and projects that once had the green light were now at a screeching halt.

The Cold Lake Project was shelved and the joint-venture team broke up. All the contractors I worked with, including the underemployed geologist Dianne Ball and the new grad Fred who looked 10 years his senior, were sent back to the agency. I never heard from Dianne and Fred again or from Hao, a new arrival from mainland China. Hao was so naïve about office politics that Dianne took it upon herself to educate him. She would get frustrated in her efforts to teach him, when all he responded back to her with a blank look in his eyes.

The Cold Lake Project went out with a reason to party. No spouses were allowed. Just workers. There wasn't the tension at the Christmas party, when everybody is trying to impress each other. It was just a chilled sit down dinner at the Calgary Convention Centre. Elevator music played in the background, as some people actually got up and danced. Booze was poured liberally, especially the wine. Hao had more than his fair share to drink and was becoming intoxicated. Roy was laughing. *Cymbidium* orchid corsages were handed out to the ladies, along with T-shirts – evidence that we worked on the Cold Lake Project. As Linda who organized the party said, it was money well spent.

Esso figured that when Trudeau would no longer be prime minister, the Cold Lake Project would be back on the block and they wanted the old team back, guessing that this would happen in about a year or two. That never happened. It would be at least five years before the Cold Lake Project would get enough steam to get going. By then, all of us would be on our merry way. Linda was extra happy. She flaunted the huge engagement ring one of the project managers gave her. Yes, she found love on the project.

I was fast approaching 30, reaching an expiry date in terms of pro-creation. A Texan engineer concerned about my life as a spinster, enlisted me to join the Calgary Ski Club. For the next two years, I looked forward to work on Mondays, to recover from weekend recreational ski trips, hikes and other outdoor activities.

One Saturday morning, I needed to go into the office and grabbed a yellow cab to work. The East Indian taxi driver was an underem-ployed electrical engineer. As I fastened my seat belt, he said repeat-edly, "It wasn't your fault. It wasn't your fault."

I snapped back, "What are you talking about? Just because I fastened my seat belt?"

His gaze ripped through my soul. He looked blankly not at my face but the periphery of my body. I got that, so I asked him, "Are you special?"

He smiled back, "Aren't we all?"

I probed him, "Do you have ESP?"

He nodded, "I'm reading your aura."

Then he added, "You were in an accident. The guy ran a red light. It wasn't your fault."

I thanked him for that. Now, I knew he wasn't full of bullshit. I lightened up. We exchanged small talk about what I did for work. He pressed me for my future, and I pushed back, "I need to find a man."

He retaliated. "No. You need to work."

I was in disbelief, "Why?"

He reiterated his message, "We need you. We need you at the top."

Only decades later, would his prophesy make sense.

I didn't volunteer to go work in the field office. It didn't feel right. It was inevitable, come December 1980, I was laid off, along with three others. There was a 50-plus grey-haired colleague, who already had a hard time finding a job because everybody told him he was "too old." We had all received a 30-day notice, and we were expected to work until our very last day. But Dave covered for me and encouraged me to get the secretaries to up-date and type up my resume and start job hunting. The efforts were probably futile. It was a month before Christmas.

My last day at work was surreal. Everybody took us out for lunch, ironically at Chi Chi's, the same Mexican restaurant that I was taken to for my second job interview. I received a farewell card, signed by everybody; even some small presents. As I returned to my cubicle to pack up my personal belongings, the beautiful technician who I thought was once friendly to me, barged into my cubicle and blurted out, "Serves you right for being an engineer. Now look where it's got you."

I was shocked by her unkind words. It was only three months earlier, that she surprised me with a pot of expensive tastefully arranged dyed dried flowers to decorate my office. Time heals. Years later, I would run into her and understand where she was coming from.

I remember at engineering school, some fellow student predicted that it wouldn't be the engineers who were against the women engineers, but their wives. Thankfully, I experienced the animosity early in my career. The company Christmas party is a corporate event that is a time for management to evaluate employees, but more importantly, their spouses. There was so much pressure at engineering school for the guys to bring dates to the annual engineering ball that the same applied to attending the annual Christmas party.

I had a date for my first Fluor Christmas party, but he called to tell me that he had the flu at 5:00 p.m. It had been arranged that our group was going to share a table together. Without considering the implications of attending solo, I came wearing a turquoise silk shirt dress, which kept me covered up to my neck and with a skirt down to mid-calf. I sat down to the left of George and his wife. Introductions went around the table. Adam's date was the effervescent secretary Connie. I was regaining my bearings, after the plated turkey dinners were being served, when out of the blue, Fiona lunged across her husband with her dinner knife pointed at me, as he struggled to restrain her, "Are you going to be working in the field with my husband?"

All heads turned on me. My mouth turned parched. All I could see was this knife in front of me. I turned to George, as he struggled with Fiona. "I don't think so."

Adam saved me, as the Christmas party had turned into an end of the world Armageddon scenario. Everyone had stopped eating. The sound of cutlery came to a halt. Adam rose from the table, tossed his napkin down, walked over to the opposite side of the table, leaned over George's shoulders and wrapped his arms around him, "Oh, Fiona, it's me you have to worry about. I swing both ways."

I think others at the table may have laughed. I ate my dinner in haste. The turkey tasted awfully dry. Just after dessert was served, I quietly disappeared into the night. I realized Dave may have been right. Christmas parties are the time when people drink too much and make fools of themselves.

I look back at working at Fluor with fond memories. I think it's because of the relationship I had with my supervisor Dave. He wasn't a stereotypical engineer of that time. Dave and his wife weren't planning to have a family, and she had her own career as a psychologist. Somehow, his marital state didn't resonate fully with the corporate credo. He had a sense of humour, and we shared a common food addiction – potato chips – which he kept a stash in his drawer and would occasionally come over to share a few with me. Dave survived another year at Fluor before he got laid off. He reminded me how lucky I was one of the first to be let go. Being left behind on a sinking ship is not a great feeling.

Although, I said I wouldn't share a place with roommates again, reality dictated another scenario. I ended up sharing a townhouse in the once TM Siddhu Village in northwest Calgary with an art college instructor and a 20-year-old pre-law university student. The rent was about a quarter of what I was used to paying. All three of us had something in common. We practiced transcendental meditation. There was a nice community in the townhouse complex, celebrated by occasional potluck dinner parties.

During the winter of 1983, I half-heartedly looked for work during the weekdays. On the weekends, I cross-country skied with the Calgary Ski Club. The skiing was great until I entered a race. The skier in front of me wiped out in the last kilometre of the course. I swerved off course and slammed into a tree. About the same time, I came to my senses about that pharmaceutical sales representative I had been dating. We parted our merry ways. In no time, I was on the telephone, crying the blues with Tony, the Canada Manpower Centre counsellor, ready to go back to work.

Chapter 25 –
Working for a Legend

I scoured bulletin boards at the Manpower Centre and read *The Calgary Herald*. There was a position advertised in the newspaper for a junior engineer, with resumes to be sent to a post office box. Tony warned me that it was common for employers to not identify themselves, so they can identify disgruntled employees, or inform their competitors that they're hiring for fear of "poaching". At that time, before LinkedIn, there were two schools of thought on resume writing – send in a comprehensive resume that listed every course you took from basket weaving to first aid and hobbies and interests; or a resume, no longer than two pages. The truth is that the legendary Lloyd Alexander, known for his unorthodox well-testing techniques, got a deluge of resumes and let his wife browse through them. She told Lloyd to hire me because I put yoga on my resume as a hobby, something she practiced, too.

Beggars can't be choosers. I took a twenty-five per cent pay cut in joining Delta-P Test, but I was just so happy to have a job. There was a downturn and even with the pay cut, I still could afford to eat, sleep, and play. I went from working for a multi-national engineering procurement company to an itsy bitsy one where Lloyd came and went as he pleased. There were also, a couple of field guys who were on call with mobile vans to run pressure tests. I was the office manager and did everything that was necessary – program the Commodore 64K personal computer in BASIC to automate field reports and plot well test curves, and sleep with a pager to mobilize field crew. It was inevitable that well tests were typically done at night, preferably after midnight. Scouts who were hired by oil companies were trolling the fields to determine how much gas was present from the size of the flare.

Lloyd spent his whole life trying to get his father's approval. When he graduated from university with his engineering degree, he told me that his father didn't congratulate him or tell him he was proud of Lloyd. Instead, his father told him, "Yeah, but I bet you can't get a job."

Lloyd got a job working for Imperial Oil. He went home and told his father, "Hey, Dad, I found a job with Imperial."

His father snorted at Lloyd, "Bet you won't last a year."

To spite his father, Lloyd worked at Imperial for 25 years Then he took early retirement to pursue his dream of having his own company.

The first thing that Lloyd got all his employees to take is the Erhard Seminar Training or "est" training. Founded by Werner Erhard, a former encyclopedia salesman from California, "est" training was held over two weekends from the 1970s to early 1980s. Celebrities testified that "est" was a life game changer.

I attended a recruitment evening at the Calgary Holiday Inn downtown (now Ramada Hotel) and was swarmed by Erhardt Seminar Training alumni, who were drunk in anticipation that I would sign on the dotted line. Lloyd paid for the tuition and airfare for two 60-hour weekends in Vancouver, but I had to pay for accommodation. I complied as I wanted to keep my job. We were locked in a hotel conference room from 8:00 p.m. to midnight the first night;

8:00 a.m. and onwards for the Saturday and Sunday. It was tough love for the two hundred attendees. We signed contracts on day one not to eat, drink water or pee on our own free will. The session started with the facilitator yelling at us, "You are assholes, every single one of you. Fucking assholes."

That would set everything in motion, and not end until 60 hours later. While I didn't think I was drinking the same Kool-Aid as everyone else, for some transformative reason, we left the second weekend euphoric about life. But returning back to Calgary, I failed to see any major breakthrough for myself. Maybe, it was more about accepting Lloyd, with all his fatal flaws, for himself.

As everybody who knew Lloyd told me, nobody lasts longer than a year working for him. Delta-P Test was a dinky operation, comprised of Lloyd, myself, and two field technicians. When he wasn't shooting the breeze with us, he spent most of his time schmoozing with his clients in the Oak Room of The Palliser Hotel; which he fondly dubbed, "The Paralyzer Hotel – as that's what happened to everyone after one drink too many."

After my gig working for Lloyd was over, I finally got a grip on the art of schmoozing. I attended a UBC Engineers Alumni reception at the Palliser Hotel. There I met some UBC engineers who graduated a few years ahead of me and were gainfully employed at Husky Energy. I followed up with a meet and greet with one of them.

Chapter 26 –
Claw Your Way to the Top

Imagine being a lobster stuck in a cage with a whole bunch of other lobsters, clawing at each other, but nowhere to go. That's kind of how some of us felt, working back in the 1980s. Humourist Dave Barry wrote *Claw Your Way to the Top*, which epitomizes the ambition and desires of baby boomers. Some of us valued the keys to the executive washroom, the corner office in the penthouse, the efficient executive assistant who did the real work and the hordes of protégés willing to sleep with them. It would be decades later, when a brief stint in management revealed to me, that the executive ranks came with strings attached.

I got hired on as a contract employee for Husky, before I got hired on a permanent basis in a new department. I secured an intermediate engineering position as Planning Engineer in the Project Planning and Development of Thermal Operations, at Husky. Then and there, I began my on and off love affair working for Husky.

Stepping into the offices of Husky, at that time, put me on Cloud 9 – after paying my dues in the cube farm. Imagine you've been renting basement suites all your life. That's like being a cube farm dweller. Then suddenly, you've been moved into the penthouse suite. That's the Husky building. Although truth be known, I never made it up to the 40th floor to talk to the CEO. He, not she, is heavily guarded by security and iron maidens. The Husky building was and still is an architectural master-piece, that has stood the test of time in Calgary with its granite exterior, revolving doors and brass accents. The foyer has this magnificent entrance. You know and feel your importance, as you step across the tiled granite floor. The sound of each step your feet makes, announces your grand arrival, as you push through the front revolving doors. Then you're hit with elevator music and the ringing of the elevators, which takes you up into the ozone layer – where egomaniacs resided.

Sure, Lloyd gave me a window office in the old Petrochemical Building a block away. However, it just wasn't the same with the linoleum floor-ing and bargain basement furniture. My very own office at Husky – floor to ceiling windows, plush carpet under my feet, oak veneered furniture and my very own office plant. Hired help came in weekly to prune, water and wipe the dust off of the leaves of my corporate issued plant. Why that's like having your personal gardener. Then every day, the housecleaning staff came to empty your garbage, vacuum, and dust. How good is that? If you're going to spend most of your waking hours somewhere, then why not work in pleasant surroundings?

Then there were the perks. Imagine a company that cares enough for their employees to have an on-site infirmary, along with its fitness centre, complete with noon hour aerobics, after work yoga, racquet courts, running track, weights, and exercise bikes. Husky also had its own corporate jet. Later, I learned but did not take full advantage, like my male peers did, was the coveted corporate expense account.

I could get used to this good life. Wearing navy to work wasn't such a bad idea. Those Colour Me Pretty fashionistas of that day pro-claimed navy, burgundy and grey was flattering to my "winter" skin tone. Finally, within six years of graduating, I was going somewhere. My ego tripped on every moment of this illusion.

Chapter 27 –
Cover Your Ass

My letter of offer from Husky was contingent on passing a medical exam, which meant a physical exam from the company doctor on staff. By now, I was corrupted working for others. There was no turning back, since I had already crossed the line. To get my job at Husky, I twisted the truth. Everybody does it, they tell me. Others will say, it's called marketing. Others call it branding. It's not to be taken too seriously, as it's just a game.

Husky's corporate medical questionnaire was extensive, more questions than my family doctor asked when it came to birth control and sexual history. I deftly skipped all the questions about contraception as technically, I wasn't married – just living in sin. So that, I told myself, was legit. After all, I had learned from dating a lawyer who was double-dipping in the dating pool, "You didn't ask. I never answered. I never lied."

Maybe, Husky could have run a survey on how they could improve their medical questionnaire. But the year was 1984. The time was pre-Internet, pre-digital and you could imagine how challenging surveys would be to implement; not to mention the countless number of trees killed in the process.

So I skipped the contraception questions as to birth control. I wonder. If a female applicant of child-bearing age put down that she was using the rhythm method, how that would affect their decision to hire. Previously, at Reid Crowther & Partners, I had seen a female colleague get demoted when she became obviously pregnant. Hearsay at the City of Calgary, married women were expected to resign upon tying the knot during the 1960s, or they'd be fired. No questions asked. I wasn't done, yet. Husky had the audacity of asking me a whole bunch of personal questions, which was really none of their frigging business, but from their legal department … it certainly was. They were looking out for themselves. What if one of their female employees took them to court for a miscarriage, due to exposure to the toxic chemicals at the refinery so freely liberated? If she had disclosed that she already miscarried previously, their backside would be covered.

So how did I answer the remaining questions? Like a virgin, of course, responding to the following questions with "Not applicable". Which reminds me? They forgot to ask me if I was a virgin. No, I never had gonorrhea or syphilis. I never had been pregnant. I never miscarried. Then, there was the clincher, intimate details on my menses. What days of the month would I be menstruating? How long did they expect my period to last? How was my flow? Was it light, moderate or heavy? Did I get cramps? OMG, they forgot to ask if I had sexual intercourse during menstruation as a means for birth control!

I took in a deep breath and sighed as I completed my answers. I thought it was best to make me look like a professional female menstruator, with dates that would coincide with pay periods. I arbitrarily wrote that my periods commenced with the second pay period of the month, the 15th of course and lasted no more than 4 to 5 days. Anything longer would concern them about my work productivity. After I finished this grueling questionnaire, I handed in the four-page forms, dated and signed them. Then I obediently peed into a plastic cup and gave them 5 ml of blood, likely to confirm that I wasn't pregnant.

It was made clear to me that the answers to the medical questionnaire were confidential. My superiors wouldn't have access. I am not so sure about this. I just imagine some supervisor tracking women calling sick one day every month due to cramps. If they ever accused me, I had my ass covered. The 15th of the month was payday. I was entitled for retail therapy, since there were no company therapists available. Why not hit those racks before others? Early birds do get the worms.

I must have passed the medical exam because I got the job. The pay disparity began from day one. Tim, the Californian born engineer explained to me that the company didn't give me credit for the non-oil and gas experience. Sigh. Now, I earned twenty-five per cent less than my male counterparts, who worked for oil companies the moment they were extruded from engineering school. I didn't care. I was elated that I had found nirvana – a job working for an oil company. Besides, I had an opportunity to claw my way to the top, like any money grabbing power hungry corporate rat.

My troubles were only about to begin. Tim made the decision to hire me for affirmative action reasons, which I would soon discover. I didn't report to Tim, but to Rick Davidson – voted by the office women to be the best-looking engineer in the office. Rick assigned me to a windowless office spitting distance from his. When I was introduced to everyone on the floor, there were numerous empty window offices, even one occupied by someone of less rank than me, a technologist. I confronted Rick on this discrepancy. He begrudgingly gave me a window office. Going from a three-window office as a contractor to a windowless office was just a reason for me that was not acceptable. It was bad enough that I took a pay cut.

Chapter 28 –
Golden Boy

As far as supervisors went, Rick was totally opposite to Dave from Fluor. I reported to duty at 8:00 a.m. sharp. There Rick stood, looking out of his massive three-window office. He gazed over the Rocky Mountains horizon and quipped, "Engineers are God's gift to the planet."

He then looked at me, as if he thought he was some sort of a sex symbol. Then Rick took another slow sip of his black coffee. The family photo on his credenza revealed a tall leggy blonde stay-at-home wife with their three young children. His photogenic wife, probably could have won a beauty contest or been a super model. Supporting a family of three was a bone of contention, as I could detect strain in living off one income. Not that I had anything against people from Saskatchewan, Rick was a mechanical engineer who got lucky. He worked on a thermal oil project that made money. Exploiting resources was the mantra of the 1970s and the 1980s. While Greenpeace was waving its flag years earlier, oil companies were complacent, possibly understanding that this was a fad.

Rick's ascent into management had nothing to do with his strong technical skills or fabulous personality. It was a matter of timing. He was just in the right place, at the right time. There was nobody else to promote. The others had been laid off in the last downturn. Rumour had it that his glory on the thermal oil project may have had something to do with the back busting work of one of his summer engineering students. I witnessed the Peter Principle in action. Everybody rises to their level of incompetence. Rick was no different.

Kevin, a multinational oil company refugee advised me, "You've got to train your boss."

There was some truth to that. He was such a weasel, getting Harvey to pay for his lunches every day with his peers from other oil companies. Kevin and Harvey became good buddies, going for lunch and beers, after work. What chance with Rick did I have? He went to the gym at lunch and straight home after work.

Chapter 29 –
What Can I Say? Isn't this Karma?

I had settled quite nicely in the two-paneled north window office overlooking downtown Calgary from the 26th floor. It was a bigger deal if you got a three-window office, but that was reserved for staff managers and above. It was just a fluke that I had occupied a three-window office as a contractor, in a different department, months before. That didn't matter. Life was going swell for me.

Over the first weekend in March 1986, I married the man of my dreams – granted an import model from the Netherlands, with his broad shoulders and soft hands to touch – a geologist who worked for another oil company. It was a small intimate wedding with my mother and close friends present, followed by a dinner and a reception. My husband, who wasn't feeling well slept most of the day, woke up at 5:00 p.m. and threw up from the stomach flu. We were married at 6:00 p.m. The next morning, I woke up and threw up, too. Even though I was sick, I was happy. Life is so much better when you're in love, whether it's for real or not and whether it's for a moment, a day, or many years to follow.

I went back to work the following Monday. My change in marital status was as uneventful of the passing of the Siamese fighting fish found belly up in my mini-aquarium on my desk. This could have been attributed to the modest diamond engagement ring that I had been wearing for several months prior. Or the fact that most of the people in the office were already married. Life went on. Then something more significant rolled around. Karma, some people call it.

Heather, the congenial secretary, excused herself temporarily and forwarded calls to my office. Of course, my phone rang. I picked it up, expecting that it was probably someone on the floor, who forgot their card key. It wasn't Carol or Kevin or Tim, likely culprits. There was hesitation. A man's voice cleared. In his low raspy English accent, Peter Kennedy, my former boss from Reid Crowther & Partners, was on the line, "Nattalia?"

I recognized his voice, "Yes, Mr. Kennedy. I will let you in now."

The tables had turned. Reid Crowther & Partners was bidding for work on the North Saskatchewan River Water Treatment Plant. Up to now, I had not known. Mr. Kennedy, looking as daft as ever, a few pounds heavier and a few hairs less on his head was completely flushed, with elevated blood pressure and rapid heartbeat. The bad girl in me got the worst in me, as I ushered him to the boardroom, "Well, I guess I will have to put in a good word for you, won't I?"

He detected sarcasm, look of horror prevailed. What can I say? What goes around, comes around.

Chapter 30 –
Meridian Avenue

The town of Lloydminster, which is due east of Edmonton, straddles the border of Alberta and Saskatchewan. In 1984, I would make my first trip to Lloydminster, which commenced my love-hate relationship with this bi-provincial town – unbeknownst to me, not end until the next century.

For the summer of 1984, I worked in Lloydminster, a town with Husky as the centre of its universe. I worked on the Aberfeldy Steamflood Project, a huff and puff cyclic steam heavy oil project, comprised of clusters of five wells grouped together for optimum production. Unlike conventional oil, heavy oil has a thicker consistency, more along the likes of molasses, and pumping alone isn't enough to bring the heavy oil to the surface for commercialization. At the Aberfeldy Steamflood site, the central well in the five well cluster was designated for steam injection. Steam would be injected downhole for a prescribed number of days. The steam would reduce the heavy oil's viscosity and make it easier to flow. Then the four other wells in the five well cluster would pump the warm less viscous heavy oil up to the surface.

Tim thought it would be a good idea to send me to Lloydminster, so I could learn firsthand what went on in the field. I would work as a battery field operator, running around all day, taking measurement at the wells – their production, pressure, temperature, and some basic lab analysis.

There was another reason to send me to the site. That was to keep an engineering student company. Cassie, a tall slender woman from Edmonton, was torn about going back to university. Her boyfriend wanted her to quit engineering school, marry him and start a family. Cassie told me that she found it hard to relate to the women who worked in the Lloydminster office. The majority of the women there worked as secretaries, married and pregnant, or married with children.

Husky supplied us with hard hats and a $75 voucher for steel-toed boots. I ordered mine from Red Wing Shoes in Winnipeg, the only place in Canada then that made women's steel-toed boots. They cost me $150, but I was grateful to wear boots that fit my small size 6 feet. Back then, hard hats and steel-toed boots were the only protective safety gear requirements.

Black leather jackets and blue jeans was the dress code amongst the visiting engineers. However, I wanted to blend in as best as I could and look like one of the operators. There were no lady coveralls being sold at Mark's Work Wearhouse. I purchased the men's size small. I separated the coveralls into top and bottom pieces, by running a pair of scissors along the seam that sewed the bottom to the top. Then I made a stretchy waistband by sewing in a piece of elastic around the waist. This separation made going to the bathroom way easier than if it was a one-piece jumpsuit.

I also knew that being an engineer was going to be used against me, in the field. (Engineers had a notorious reputation of making the field staff miserable, for ignoring their suggestions for improvement.) Before I entered the field trailer, I slipped my iron ring off my right pinkie finger. Even if I told the field staff that I was an engineer, I figured that they would never believe me. Better to think of me, as a non-threatening technician.

Husky flew a corporate jet into Lloydminster for the first five or seven passengers every day. The Cessna stuffed everybody in like sardines. Passengers enjoyed complimentary watered down liquor and pop, stashed away under the seats. If you missed getting on the plane, you had to fly Time Air to Lloydminster in refurbished military planes that somebody in the office nicknamed, "Tin Can Air".

No liquor was served, but they had a flight attendant who poured coffee and beverages. I never thought much about flying Time Air, until we hit turbulence. The cockpit door flung open. The juice in my plastic cup leaped a foot above into the air and splashed down all over my right arm. The consultant beside me was battling her fear of flying. If I'd known it would have been this bad, I may have suggested that she drive to Lloydminster the day before.

My fellow male colleagues teased me about the tiny bladder syndrome that inflicted women. That was the hardest part of my job. There was no toilet on the Cessna or at the Lloydminster Airport. I sought refuge at our inaugural stop at Tim Hortons. That's why I volunteered to buy a box of donuts, making sure that there would lots of chocolate long johns and few jelly doughnuts. After that, I would have to hold my bladder for the next four hours, until lunch. This task was not entirely impossible, but very hard when I was pregnant. Now, if I could only pee like one of the guys – standing up, at the back of the truck or in a bush. Back then, the smaller field operations known as batteries, for collecting and holding the heavy oil, never had washroom facilities.

What provided me relief was the Aberfeldy Steamflood site. That was large enough to have an office which included one gender nonspecific washroom, microwave and running water. The toilet seat was always up, but this wasn't a battle worth fighting. I'd pull off a few sheets of toilet paper to put the toilet seat down. After I was finished, I flipped the seat up. There was no reason to tear down the Playboy pinup on the wall. Another one would be in some operator's truck.

I thought it was good enough of me to bring doughnuts to the site office, but there's no friggin' way I was going to pour coffee. While I never had a plan for my career, I did have a plan not to serve coffee. No negotiating. I was prepared. I always brought a thermos filled with decaf coffee. There was no need to share. Mention the word decaf. That's insinuating you've got herpes or some highly contagious disease.

I was this scrawny brat trying to get my weight to tip over one hundred pounds. I could chow down like any other guy I worked with. Albertan boys liked their meat, especially beef, and best viewed for public display as a steak sandwich for lunch. Not fearing for my cholesterol levels (yet), I had the gall of ordering the most expensive luncheon special at the Triumph Restaurant located next to the Wayside Inn (now Days Hotel) on 44th Street – the three lamb loin chops grilled to perfection with rice and salad. I knew what they're thinking to themselves. They were waiting for me to say that I was full half way through the meal. That never happened. It must be the country air. I was always famished at meal times.

Cassie and I worked a 12-hour shift at the Aberfeldy Steamflood site. We rose when it was dark outside, drove to the Husky Truck Stop for breakfast, and shoveled our faces full of scrambled eggs, pan-fried potatoes and bacon. When in Lloydminster, I didn't want to look like a pussy. All the people were heavy on the meat and carbs and so, a city girl like me had to man up.

Field work was a break from the mundane existence of working in an office. While Cassie and I took turns driving the field truck to Aberfeldy, I had a rule of declining to drive in the presence of other male colleagues. I feared any criticism about being a lady driver and any puns that could be made about me being Asian.

Meeting etiquette was something that I flunked. Rick asked me to take notes at a meeting. I flatly refused. Not in a gracious way, I can add. I kept my wrath under control, which manifested in the development of temporomandibular joint syndrome (TMJ). The eyes were rolling in the room, as Rick looked to Tim for backup. There was this awkward silence. Finally, Tim asked me to take notes. I complied, feeling that I had lost a major battle. My two contemporary engineers, Murray and Kevin, witnessed the boardroom skirmish. They discussed the incident with Harvey who agreed that we, the intermediate ranked engineers, should take turns in recording meeting minutes.

While I was the first woman engineer hired into Project Planning and Development, there were other women engineers at Husky, too. Sara, an attractive blonde, was located in the Lloydminster Office. She confided with me that her peers took pleasure in targeting her body parts with rubber bands. I'd admire for her composure when a drunken co-worker broke into her second story apartment at 3:00 a.m. The married man professed his love for her. She calmly called a taxi and sent him home. Nobody got hurt. Not even his wife.

Chapter 31 –
Behind Closed Doors

Girlie magazines freely circulated inter-office mail. Sex in the boardroom was not entirely impossible. Marcel, a male colleague, who had some oversized drawings to review, accidentally walked in a boardroom catching his female boss and another colleague with their pants down. This incident happened one weekday mid-afternoon. No words were spoken at that time. Marcel was in a vulnerable position. No different than a male in authority, Marcel's boss began harassing him. First, he confided with me that she would rub her body a little too close for him while he was at the drafting table. Wearing a V-neck top without a bra, she bent over the table, exposing herself. He was not impressed.

She persisted, until finally, he came up with a comeback at a corporate function. She greeted him, "How are you doing, Marcel?"

He approached her, and he told me, "I grabbed her butt and said I'm doing fine."

Then he grabbed her breasts, "So how are you doing, Monica?" She didn't know what hit her. From that point onwards, she reverted back to more professional behaviour around Marcel.

Seinfeld and *Dallas* were the television shows that everybody at work watched. But Dallas resonated unanimously with everyone in the boardroom. The CBS television saga of the Ewing Family – a bunch of backstabbing, money hungry, greedy and lusty bastards with nepotism at every angle, aired every Friday at 8:00 p.m. On Mondays, around the morning coffee station, conversations revolved around what happened to J.R.

Some people think there's some truth about *Dallas* as what happened behind closed doors for Calgary blue-eyed sheiks. It was common practice that if you got your friend hired, you would get rewarded with a finder's fee. Vendor customer appreciation involved going for lunch at The French Maid, a local strip bar, which was across the street from Delta-P Test Office. Lloyd often would comment with a grin on his face about the walk of shame, as working stiffs left the joint. No wonder they didn't want women working in the oil patch.

Chapter 32 –
Valentine's Day Massacre

After 1980, reduced demand and increased production produced a glut on the world market. The result was a six-year decline in the price of oil, which culminated by plunging more than half in 1986 alone. The inflation-adjusted real 2004 dollar value of oil fell from an average of U.S.$78.72 per barrel in 1981, to an average of U.S.$26.80 per barrel in 1986. We all thought we were invincible. Our egos were on steroids. From the 26th floor, we saw God's country every day. We thought that we were gods and goddesses. Nothing could be further from the truth.

People in the know were telling others to start packing. I thought they were being over reactive.

A memo was issued a day before, announcing that Husky would lay off three hundred persons over a three-day period. At that time, I was four and half months pregnant with our first child. I wasn't planning to tell management that I was pregnant, until it was obvious. After the memo was issued, I walked into Tim's office, thinking that my pregnancy would grant me immunity. He nodded when I told him. He calmly told me, "Don't worry."

I went home thinking like Fluor, that Husky wouldn't lay off any pregnant women. I was wrong. The next day, I went to work, believing that

I wasn't going to be slaughtered. Shortly upon my arrival, Rick walked in my office to give me my termination papers. I received two month's severance pay for a little more than two years of service. My intent was to work for at least five years before switching jobs. That certainly wasn't going to happen now, was it? As soon as Rick left the office, I shut the door and let the tears overtake me. I ran my hands over the fine oak-veneered desk, which I so enjoyed working on. Now that dream had just turned to dust.

I will give them credit for letting me stay at the office, as long as I wanted. It was okay to have walked out then and there. I came in the office for another day and then excused myself. I wasn't alone. Sara in Lloydminster got laid off, too. The phones were ringing off the hook to find out who got laid off and who remained. Most oil companies did not let too many people go in the last downturn, a few years earlier. But 1986 was different. When the price of oil drops from USD$28 to USD$14 per barrel overnight, the hemorrhaging was going to be bad.

I heard later, that in Kevin's group, Harvey called everybody at home at 7:00 a.m. that morning. He re-assured everybody that nobody in his group was going to lose their jobs. Everyone was advised to come in later that morning. In fact, nobody in Harvey's group showed up before noon. If they did, they were probably told to sit in their office and keep their door closed.

There was no going away party for us voted off the island. Instead, someone called everyone for a dinner party at The Spaghetti Factory on the third day of the lay-offs, March 12, 1986. Some said they were going to travel, others married to other professionals weren't on financial life support and for the most part, or maybe it was the liquor or denial, but that evening, we had a good time and vowed to keep in touch. Most of us never did. There would be another boom, we hoped, as bumper stickers said, "Dear Lord, please give us another boom and this time, we won't piss it away."

PART THREE –
AFTER THE BUST

Chapter 33 –
Sweat, Motherhood, and Tears

Four and half months after I was laid off from Husky, Andre – our first child was born. I kept hoping that he wouldn't be late. I prayed that his head would not be the size of a 10-pound bowling ball because I anticipated that would hurt like hell. I stuffed my mind reading all the pre-natal books that ever made it into print. I went into the birthing room at Grace Hospital to psyche myself up, as hospital smells always made me queasy.

Can anyone be really prepared for parenthood? Would I be like that fourth time mom in the birthing video who doesn't sweat during labour? Why, her mascara doesn't even run! If that pregnancy glow isn't a bronzer on her face, what is it? I never glowed when I was pregnant. I felt like I had PMS for nine months. My clothes didn't fit. Silk and wool yielded to polyester knit. My swollen tits were sore for nine months. My belly was ginormous, and I got very cranky, especially, towards the end of the last trimester.

I won't bore you with details about my labour, as it's obvious that this normal natural process has occurred several billion times before. You already know the world's population is over several billion by now. I got my wish. I went into labour two weeks ahead of my anticipated due date. At 5:00 a.m. in the morning, Andre arrived – a five and half pound bundle of poop, pee and joy, with eyes slammed shut. The little bugger was obviously not a morning person. Neither was I. But for this occasion, I was.

Just like they said in the Lamaze pre-natal classes, he would have ten fingers, ten toes and feel as mushy as freshly-baked white Italian rolls. If I could classify Andre's baby status, I would call him a polyester wash and wear one. He wasn't normal. He slept through the night within a few weeks. He sat in a soiled diaper for hours without complaining. He only cried when he was hungry or tired, and he didn't mind listening to Vivaldi's *Four Seasons*. He was reserved, compared to the other babies in the pre-natal class, as his eyes absorbed everything around him in one gulp.

Motherhood was turning out better than expected. Andre took his naps without a fight and slept through the night like he was in Heaven. Being laid off was a blessing. I even found time to sketch. The thought of working as an engineer turned to Pablum. I wasn't alone. The 1986 oil bust threw thousands of engineers onto the streets.

It was a sunny day in early December 1986 at 9:30 a.m. – Andre had just been fed. I changed his slightly wet diaper. Our eyes locked longer than usual. I swore he was smiling at me. I snapped the buttons on the powder blue sleeper he wore, hugged and kissed him for his morning nap. I went downstairs to do some laundry.

10:30 a.m. – I expected to hear Andre's cry, that he was ready to get up and go for a walk. Hmm, he's not up. That's okay, I told myself. He was crabby the night before. Four months, nine days old. He's probably teething. Those two lower front teeth.

11:00 a.m. – I was overcome by this sinking gut feeling. I flung the nursery room door open. The silence was frightening. I bolted towards the crib. There he was. Motionless. He lay on his stomach. His head rested to the left side. Andre's right fist had yanked the blanket partially, over his head. His left fist clenched in defiance by his left side. I became the epicentre of a seismic shit show, 10 on my Richter scale. I trembled as I grabbed my baby's still lukewarm body, clutching him to my breast like my life depended on it. His eyes didn't open. He wasn't crying. I panicked.

I placed him on the living room sofa, and kneeling on my knees, I applied artificial resuscitation. I blew into his tiny lungs so hard that his stomach threw up. I called 911, shrieking hysterically in an incoherent voice, "Please help … please help … my baby's not breathing."

Everything that follows is a haze. The seconds seemed like light-years, as I paced in a circle around the house.

When the emergency responders arrived, I eyed the oxygen pack one paramedic carried. I expected him to hook Andre to the machine right away. Damn it! Goddamn it! No, fuck this. What the fuck is the wrong with you guys? I called you to make my baby breathe again. He's not breathing. He needs to breathe. Are you just going to stand there? I was hysterical. They just stood there. Long blank faces. It's like they were seeing ghosts. I felt robbed. I was cradling Andre in my arms, when one of the two paramedics gathered enough courage and said, "I'm sorry, your baby is dead."

My jaw dropped. I looked down at Andre to see his lifeless face. I freaked out that he was dead and handed Andre to the paramedic, "Please take him."

My voice tightened. I couldn't utter any words. My mouth was as dry as the Sahara. My head throbbed. I retreated to the kitchen from the living room. I sat on a chair and crouched over in a foetal position. My world caved in. I was sick to my stomach. I didn't know what to think. I didn't know what to do. I didn't know what to fe' The grief inside me was locked up in Fort Knox. The cops arri' with my husband. Now, I felt like a criminal. Doubts imploded in my mind. Was I bad person? What did I do wrong?

The cause was sudden infant death syndrome. Back then, that was one in 500 births.

I never imagined that this would happen to us. Andre was planned for, wanted and cared for. One thing became perfectly clear for me was my priority in life. During my pregnancy, we were preoccupied in hiring a nanny, as it was assumed that I would return to work after maternity leave. The silence in the house was unbearable. All we could think about was having another baby, without fully comprehending the grieving process of losing a baby. I couldn't bear the words of what another SIDS mother told me. This son, our baby Andre, will impact us for the rest of our lives. Every birthday of his that rolls by. The anniversary when he died. The annual depression that one goes into. Each year, it's just a tad lessened but still rears its ugly face when things seem to go well.

It would take me decades later until Andre's impact on our lives, would be fully realized. There is no easy answer to deal with the situation. Our grief was so heart wrenching that we chose to deny it. My husband chose to drown his sorrows. I don't know why I didn't cry, like I did at that second-year summer engineering job interview. Instead, I shut down. I just buried all my hurt and anger, until years later. Only then, I was ready to deal with such a loss. Sooner or later, I had to feel my anger, that incensed me to the core ... the feelings of guilt, shame and failure all rolled into one tsunami. Sometimes, life throws you a curve ball.

While many women engineers of my generation were torn between career and family, I now knew in a flash what my priority was. It wasn't engineering. How silly was I to think that a career should come first? I heard that men fret once they are retired, why didn't they spend more time with their children growing up. Years later, I would meet a female judge in Toronto at a *Who Who's in Canada Women* networking event, who was married with four children. If she had to do it all over again, she told me that she would have liked to have had more children; possibly, two more. Madeline Albright, the U.S. Secretary State got married and had her four children young, before returning back to the workforce full-time. She says women can have it all, career and family, but not all at once. Women have choice. Barbara Walters, the famous news correspondent says women have to choose one or the other. She chose career.

With four job layoffs under my belt, my engineering career wasn't really something to write home about. Andre's death played havoc with my self-esteem. A woman engineer called me on Christmas Eve 1986, "Did he die because you were too old to be a mother?"

I was 33 years old. Nine months earlier, I was resting on my laurels in Cloud 9. I had it all – the career, my Prince Charming and a baby on the way. Now, I was without a job and with an empty cradle. We rushed out to have our next child who was born 15 months later. Our daughter Alisse, was everything her older brother wasn't. She was loud, curious, and full of piss and joy. However, the first four months of her life were nerve-wracking for me. I would repeatedly poke her in her ribs when she slept, just so I could hear her cry.

Of course, staying home playing googly-eyes with Alisse wasn't going to serve us very well. Thoughts of returning to engineering were erased in the way a drawing with Etch-A-Sketch disappears. Even with Alisse, I wouldn't talk about Andre in public. The pain was too much. I thanked God or the powers that be that perhaps Andre's gift to me was to wake me up about the proverbial conflict between women's work and family. Working at Husky was an illusion. When I left, nobody bothered to keep in touch with me. Another woman engineer lamented, "The moment you're gone from the office, nobody cares about you. They even forget to water your plants when you go on vacation."

It's just a job. They don't read resumes at funerals.

Chapter 34 –
Write On

I turned my back on engineering. I couldn't face the people I used to work with at Husky. I never felt so ashamed of what happened. My husband said he didn't blame me for what happened. On some level, I did. What kind of mother was I? What kind of person was I? I bet my former co-workers were talking about me, judging me, rationalizing that I deserved what happened to me. I was probably wrong, but that's how I felt. They were probably numb, in shock, or just indifferent, as they seemed like the kind of people who did not want to get involved.

I did what I always did when confronted with a problem. I ran. This time, I ran from engineering. I didn't even bother looking for work. Fate drew me to entrepreneurship. University of Calgary offered New Venture Forums, which included complimentary food and booze. Nobody was judged when everybody went around the room and introduced themselves, their hopes and inspiration. The monthly speakers shared some aspect of starting your own business. I felt connected. I wasn't alone. Many of us had been fired several times. Best of all, nobody knew my past, especially, the part of being a mother of a SIDS baby.

Some of these entrepreneurs had great ideas, but they had no budget for publicity and promotion. It became my mission to get the word out on their business, invention, or product. This would be a win-win scenario. One problem though. I knew nothing about journalism. I asked for help from Steve Lane, an engineer from CH Synfuels, who I once applied to for work. His photographer wife Heather Wilson worked for *The Calgary Herald*. I asked Heather what the protocol for writing for *The Calgary Herald* was.

The year was 1988. Print media was still at the top of the media pyramid. You went to Radio Shack (now known as The Source) for all of your personal computing needs. Better still, was that writers were hired to write articles. *The Calgary Herald* was launching a new weekend magazine. I mailed a query letter to its editor Gordon Cope, who was on the prowl for freelance writers. He promptly called me up and assigned me a feature story on what I proposed – Calgary entrepreneurs – on speculation, as I did not have any "tear sheets" or for industry outsiders – writing samples to show him. The length was 1500 words. For me, this was a big deal.

My lead into the feature story was Carolyn Jones, a young unassuming artist and designer who made her greeting cards stand out with bling she glued onto premium quality paper stock. Her studio was a cramped efficient bachelor suite in Lower Mount Royal. She sold her cards on consignment at local retailers. Sales were so brisk that Hallmark card representatives demanded that store owners remove her cards. They gleefully hid her cards, when the big boys came in. Hallmark tried to buy her out, offering her consolation money for her witty words of wisdom. She held out as long as she could. The last thing I heard, I believe they offered a deal she couldn't refuse.

I wrote my feature as an engineer would – like a technical report. Mr. Cope called me up and wondered where I studied journalism. Guilty as charged. I confessed. He ranted at me for ten minutes. I feared that my misrepresentation as a writer was going to get me arrested. I volunteered to do a re-write, but Mr. Cope told me that wouldn't be necessary. When the article was published, I was pleasantly surprised that it made cover story. Mr. Cope's editing was not the lobotomy I expected. His editing was more like an appendectomy, as he removed some excess paragraphs and rearranged the structure.

After my first foray into journalism, maybe I should consider the two-year diploma course at Mount Royal College? I met with one of the journalism instructors there. He observed that many of his graduates never saw their by-line in print. After all, "the only way to write is to write more."

That I did. Except now, I pressed rewind and moved ahead crawling on my hands and knees. I called up Roberta Walker, editor for *Alberta Business* magazine and asked if she needed help. She could always use fillers, snack-sized articles between 100 to 150 words, but no more than 200 words. Of course, as a freelancer, the story would be written on speculation. My second article was a profile on Louis Stack, founder of Pro-Fitter Enterprises. A ski lover, he invented an exercise device that helped skiers work on their glutes and balance, all year round. Over the next 40 years, he would build his company to become a Calgary success story. I got paid $25 for this story, which took me six hours to write – an excruciating exercise to mince and slice my words to cram everything I wanted to say in 200 words or less. Ms. Walker was pleased. I would write many more articles to come, gradually working my way into writing feature stories. John Dodd, an Edmonton-based writer later became editor. When I got promoted to Associate Editor, I garnered the prime real estate in print of every ego – the back page column.

Chapter 35 –
If You Work for a Jerk

My dysfunctional work history was perfect for my new assignment. I wrote on many topics. None were as infamous as If You Work for a Jerk – Part I; followed by If You Work for a Jerk – Part II, circa the late 1990s. Someone told me that my back page column was posted at a Petro-Canada (now Suncor) office coffee station, and then quickly torn down in haste, as being inappropriate. What can I say? I don't want to sound conceited, but revenge was sweet and sort of therapeutic, too. The following excerpts are meant for amusement purposes and, by no means to be misconstrued, as sound career counselling advice.

IF YOU WORK for a JERK (Part I)

"Corporate rats race up the career ladder to stop watches that tell them how successfully they've made their bosses look good. But even trained rats can't succeed if they work, as so many do, for a real turkey. Like the ones who read the newspaper every morning with their eyes closed. We're talking about bosses who regularly use four letter words that start with 'f' to blame you for what they did wrong."

If YOU WORK for a JERK (Part II)

"You've got to train your boss," some weasel at work told me once. "You can," he claimed, "insult your boss in a hundred ways and still be his best friend."

That was fine for him. His boss had a sense of humour. Mine was the original macho male who still thought he was God's gift to the planet, if not females.

Forget talent, hard work and lucky breaks. How you get along with your boss can make or break your career. Will your boss talk you up for raises, groom you for a promotion – or is your success a threat to him or her? In the latter scenario, you might as well just get out, to another company or at least another department.

Your spouse, lover or family sees less of you than your boss. He can be a pit bull terrier, St. Bernard or poodle. And what chance do you have to size him up in advance before you take the job? You get a 20-minute interview, during which you're trying hard to sell yourself. First impressions are usually bunk.

One former colleague confided to me in the ladies' room. "My boss is such a jerk," she said. "But if I get another job, I'll probably get stuck with an even bigger jerk." She reapplied her lipstick and concluded: "I might as well stay here."

Chapter 36 –
Women Engineers –
Dismantling the Myths

Reporting on women engineers also became my forte. Controversy reigned during the 1980s. Increasing numbers of women engineering graduates were populating corporate ranks and upsetting the status quo for both sexes. If the choice of my subject matter branded me a feminist, so be it. Even so, I wonder why today the media hasn't blown its horn on the phenomena that only seven per cent of nursing graduates are male in North America.

Can women engineers have it all? That answer seems evasive and elusive simultaneously. In 1988, I heard the late Claudette MacKay-Lassonde speak at the Women in Science, Engineering and Technology Conference in Calgary. She was a feisty outspoken attractive woman. I believed she succeeded in having it all – a career, loving husband and two wonderful children. She was relentless in her mission to recruit, mentor, and promote other women engineers. At that time, I failed to ask her one question when I interviewed her for an article, "Was she really happy?" She certainly was highly stressed.

A year later, feminists had their field day, after the tragedy at Ecole Polytechnique on December 6, 1989. That's when Marc Lepine walked into a classroom and killed 14 women, of whom 12 were women engineering students and injured another four men and 10 women. A girlfriend of mine who worked her way into management downtown called to tell me she was very upset. His suicide letter cited anti-feminism motives. I was shocked, and horrified, like I was the one shot in the stomach. It could have been me. Until now, it never occurred to me that violence could be an outcome for choosing to become an engineer.

The media glare shone on the engineering profession in a most unbecoming manner. How did engineers who were such rabble rousers and risk takers during university transform into conservative people, who coloured between the lines? Ironically, the engineering profession, riddled with individuals who shy away from politics, human rights or leadership was now thrown into the limelight.

In the late 1980s, I conducted my survey on Alberta women engineers for an article I wrote for *Alberta Business*. They spoke to me in confidence, only if their names or employers were not identified. It helped to tell them that I had previously worked as an engineer. The majority of the women engineers surveyed were under 35 and had not dealt with child-bearing issues.

The higher up the corporate rungs the women had climbed, the more they toed the company line during media interviews. A seasoned woman engineer, who had a spot on the penthouse floor of a downtown oil company dare not take more than a couple months of work off after giving birth. She claimed that she worked 70 to 80 hours a week, and had no time for extra-curricular activities. Her supportive spouse, who gladly gave up his career for hers, didn't mind taking the lead on the parenting front. Her rapid rise up the corporate rungs had something to do with the company's optics. The president announced in the press that he very badly wanted a woman's face in the annual report. She had the right image – a balance of beauty and brains.

I wrote a back page column published by *The Globe and Mail* in March 1992 on my theory as to why there weren't more women engineers, a pet peeve of mine that persists to this very day. Today, social justice warriors today insist that for engineering to be liberated, half of the professionals must be female; the other half male. Yet nobody whines about the under-representation of men in clearly female-dominated professions, such as nursing. My article was republished several more times – in a high school book entitled *Health Issues*, and then again, used in English high school exams. Some excerpts follow below:

> "I am amused. Take Catherine or Nicole, the two women engineers being portrayed in the Secret deodorant commercial. The first is a dragon lady who is going to the top of a construction firm. The second yanks off the cover of a flashy new car to the approval of senior executives. Not a hair out of place, not a wrinkle to be seen, these women are tough but cool.
>
> While educators point to a traditional macho-male-engineer stereotype as one reason there are so women engineers, what about the negative female-engineer-stereotype?
>
> Women account for slightly more than 3 per cent of the engineers in the workplace and make up 14.5 per cent of first year engineering students last year (1991). With so few women in engineering, most people quickly conclude that for women to succeed, they must be like Catherine or Nicole. There's the expectation shared by many women who are still pioneers in the profession that "a girl must work twice as hard to keep the same ground

as the guys." Then there's the added responsibility to be a future role model – so please don't screw things up for women who may follow – even if it means sacrificing one's personal life.

These perceptions were very real when I graduated the from the University of British Columbia in engineering in 1978, when women accounted for less than 0.5 per cent of the engineers in the workplace and 6 per cent of engineering students. Such attitudes lingered in the workplace. In 1984, one engineer I worked for told me, "I think it's all right for girls to become engineers as long as they are young and single."

Despite the barriers, some women engineers have done remarkably well in the oil patch, rising up the corporate rungs as fast as they could climb. But many of us slipped off at a rate two to four times faster than our male counterparts – not after a year or two, but at the five- to 10-year mark when one simply realizes one isn't going anywhere ... One woman asked, "Why is that when a woman is assertive, they call her a bitch? When a man is assertive, he's seen as aggressive and that is viewed positively."

But be a wimp and one will be victimized in the workplace. There's a competitive world out there that, perhaps, young girls aren't conditioned to deal with ...

I've been tough, but not cool. I swore in the boardroom when I was asked to take notes. That, I admit, was not very ladylike. While young women may choose to be intimidated by the macho-male-engineer stereotype, many also choose to be intimidated by the tough-woman-engineer stereotype. Mothers of daughters lamented, "Who would want to marry a lady engineer?" (I say, practical men looking for someone to split the mortgage or help diversify their portfolios.)

When I got married, the engineers I worked with were in disbelief. I was, in their minds, a right wing Yuppie, obsessed with work and money. When we had children, my male peers nearly had heart attacks.

While the engineering profession direly needs more women, the workplace with its historical and ingrained corporate cultural beliefs has had only varying degrees of success in integrating them. Common sense, intuition, and a sense of humour really help. But being such a minority, women engineers have to be strong and many young girls prefer not to be seen this way."

Chapter 37 –
Between the Lines

Seeking more challenging writing assignments, I visited John Howse, *Maclean's* magazine Western Canada's Bureau Chief in Calgary. A tall gracious man, he worked in a tiny office crammed with magazines and books – ironically, in the Petrochemical Building on 8th Avenue S.W., the same building as Delta-P Test, but two floors up. Mr. Howse was an established respectable journalist that any young writer would aspire to become.

Mr. Howse tipped me off that *CBC Newsworld* was doing a one-hour special on *Sexual Harassment in the Workplace,* and I should indulge in some shameless self-promotion. I wasn't so sure about stepping out of my comfort zone. Mr. Howse argued that I really had nothing to lose, "It would be exciting to be part of a panel discussion."

It would be a leap of faith, for another misadventure. All my friends were so excited for me, forecasting overnight celebrity success. They were especially curious as what CBC's makeup artists could do for me.

I trekked down to the CBC Studio, coincidentally a 20-minute walk from where we resided. Why was I so hung up on putting on a pair of ladder-free stockings and my finest pair of Arnold Churgin leather pumps? What was all the excitement about? The camera only shot my face mainly from the shoulders up. I was shown to a desk to sit at, with a false city portrait behind me. Now I thought. I could have worn a pair of comfy runners. The Barbara Walters' clone and other VIPs part of the panel were based in Toronto, the centre of Canada's universe. What? I never even got to shake their hands, determine if they wore N° 5 CHANEL or exchange business cards.

I was being hooked up remotely. It was just me and the cameraman on duty that day. I wore a bright turquoise shirtdress, but my face was bare. Guess what? It was Saturday. The makeup crew were not at work. I sauntered into the dressing room, gasped at what looked like a full-service beauty counter with everything out of its packages. The problem is that I didn't know what to use. The cameraman yelled at me, "Will you hurry up? We're on in five."

I applied a bright shade of lipstick and returned to the newsroom. The cameraman shook his head. He grabbed the unisex face powder and plastered it over my face. Voila, ten seconds. I was ready.

It was an hour broadcast. Maybe I got to speak ten minutes. Most likely five minutes. Tops. The phones were ringing off the hook. One was a very angry male engineer from Ontario. My presence alone unsettled him immensely, extrapolating that I was sexually harassed and conceding, all women engineers were sexually harassed. I carefully chose my words, and explained to him that I had not been sexually assaulted nor experienced sexual harassment.

Then again, I have always been a slow learner in this department. When working at Husky, one of my male counterparts came into my office to apologize for harassing me. I was surprised. I hadn't a clue what he was talking about. Now, the story of a woman engineer's boss who rubbed her inner thighs with his hand, while she stood at the filing cabinet – that's a different story altogether. Or the female technologist, who goes to the field, is sexually assaulted and quits, while the perpetuator gets promoted? Seriously, this stuff did happen.

CBC Newsworld produced a clip, asking women on Toronto streets if they've experienced sexual harassment. A young long-haired blonde with black mascara-ringed eyes who worked as a secretary said, "Sexual harassment? Do you mean do I sleep with my boss?"

She tossed her head back like a wild filly and flatly stated, "Do I want to keep my job? Yes, I do. Do I have to sleep with my boss?"

She looked head on into the camera lens and flatly stated, "Of course, I sleep with my boss. I want to keep my job."

After my cameo appearance on national television, I thanked Mr. Howse profusely. (Too bad there wasn't social media back then. I wondered if my Twitter followers would have spiked.) Mr. Howse, then suggested that I visit Christopher Donville who was working for *The Globe and Mail* in Calgary. My timing was perfect. *The Globe and Mail* wanted to increase its presence in western Canada by publishing advertorial-driven special regional reports. While I was given some journalistic independence, I would be carefully instructed by editors in Toronto who I needed to talk to for the assignments, and the tone of the articles.

I also worked a few months for *The Calgary Sun*, in its Travel newsroom. I and another reporter, Laurie West, a recent journalism school grad, would report for duty by 3:00 p.m. every Wednesday. The Travel editor would instruct us what advertorial driven stories needed to be published. Then for the next two hours, we churned out travel stories, imagining what these destinations were like because in most cases, neither of us had ever been there. This was not real journalism, Laurie quipped. We should really travel to the places we were writing about.

The only travel opportunity we got was an Alberta bus tour. For sure, I would leave the tour, a confirmed redneck. But no, I could not bring myself to eat "prairie oysters" (young bull's testicles). These edible morsels were presented to us on the end of toothpicks to be dipped into ketchup – an appetizer at a genuine country bar that served drinks from Styrofoam cups and dinner on paper plates with plastic cutlery.

I was keen to write, or as my Grade 5 classmates would call me, "a suck hole." My search for the next writing assignment took me to the doorsteps of *Alberta Report*. This right-wing weekly newsmagazine started by Ted and Virginia Byfield that birthed many politically charged Canadian writers, like Ezra Levant. Unless you were a Carleton University journalism grad, Laurie said it was futile to get a regular gig for *The Calgary Herald*. Some Alberta writers turned their noses at writing for *Alberta Report*. I didn't care. I just wanted to write.

Sons Link and Mike Byfield worked for *Alberta Report*, in Calgary. They were open and honest with me. They didn't lie to me about them winning local popularity contests – or the continual stream of pending defamation lawsuits or the amount of hate mail they received. The Byfield family was also under financial duress. Ted and Virginia mortgaged their house four times to keep the newsmagazine afloat. Now that's what I call commitment.

The Byfield brothers interviewed me for a proposed four-week advertorial written special report to be called *Energy and the Environment*. The oil industry was starting to feel the wrath of the environmental activists in Alberta – Vivian Pharis from the Alberta Wilderness Association, the late Martha Kostuch, a veterinarian from Rocky Mountain House and Rob MacIntosh, founder of the Pembina Institute, for starters. The Canadian Association of Petroleum Producers (CAPP) was pinched with a public relations problem

– how to convince the public that the oil industry was developing oil and gas resources with a social, moral and environmental agenda. Link and Mike proposed to CAPP that they publish a special report for them. There was one caveat and CAPP wanted a fresh face to report on environmental matters. I was perfect. I was three months pregnant when I accepted the assignment. I never mentioned this to Link or Mike, as all the work was to be done way before I hit the third trimester.

Of course, that's not what happened. It took forever to get the advertising sold. I couldn't start writing until I was told who in the oil patch I was supposed to interview. I was five months pregnant, second pregnancy and I just popped out sooner than expected. Mike took notice, but didn't say anything at our weekly meetings. Four weeks later, he could not contain his curiosity, and blurted out, "Are you pregnant?"

I kept my composure, "Yes, I am."

There was such relief on his face, "I thought you were just getting fat, maybe you were growing a tumour or something."

I asked, "Is there a problem?"

Mike asked, "When is the baby due?"

I replied, "End of February."

He sighed, "We've got lots of time."

Though somewhat candid and blunt, and perhaps, misunderstood – both Link and Mike were good to work for. I was relieved that being pregnant was not being used against me. Quite contrary. In fact, Mike was quite excited. Months later, after Alisse was born, he made a surprise house call to see our drooling snot dripping baby. I just didn't see that coming.

As a freelance writer, I was guaranteed that I met interesting people. When asked about my career transition, I quipped, "Engineering is a respectable profession to leave."

Being a woman, I always had the prerogative to change my mind. I paid my annual dues for the Alberta provincial engineering association and attended technical talks to keep abreast.

When my back-up-backup to childcare had to be implemented, I brought Alisse downtown with me to the media interviews. It seemed natural for her to behave well in public. On one occasion, I was pushing Alisse downtown in a stroller, and I ran into one of my go to engineer friends, George Shaw. I don't know why, but most engineers had just three questions that they needed answered when they saw me with Alisse.

They would undeniably ask, "Is this your baby?"

I would nod and smile. "Did you give birth naturally or by Caesarean?"

I replied curtly, "Naturally."

Then after digesting this data, inevitably, their third question was, "Are you breastfeeding?"

I knew what they were thinking. How could this flat-chested woman produce enough milk to feed her baby. I gloated at them, "Yes. I'm breastfeeding".

Then I watch their eyeballs roll back into their sockets. With nothing more to say, we'd part our merry ways.

Within a relatively short time, my by-line was becoming more frequent. I wrote for *Oilweek* magazine feature stories and was a guest columnist for their back page column, too. On the Monday, before I went into labour with Cody, I accepted an editorial assignment due five days later – assuming that the doctors are never sure about due dates. Most babies seem to arrive late, just to piss off their parents.

Tuesday morning, Cody was born after a fantasy labour. In my hospital garb, I called from a pay phone to conduct my interview. I was sent home on Thursday and had my column submitted before noon on Friday. In hindsight, I should have never taken the assignment. Six weeks later, hell broke loose when my hormones crashed. I felt like a truck had run me over. Didn't know that was coming, either.

Chapter 38 –
Mad about Yellow

Decades before the Internet came of age, a Vancouver-based futurist, the late Frank Ogden who dubbed himself *Dr. Tomorrow*, spent decades worshipping Japanese technological advancements. He lived on a houseboat in Coal Harbour, likely the first Canadian to be served his morning coffee by his robot. He travelled to Japan frequently. He spoke of a world in the future, where man and technology will be integrated seamlessly and our education system, would be no match, for the speed at which our world changes. The consultant, broadcaster, LSD therapist, teacher and author of 22 books, was also a media whore who was an ideal legit source to quote profusely. His passion for all things Japanese was highly infectious. I was warming up to eating sushi. That was the first symptom of yellow fever.

In the next stage of yellow fever, I developed yellow people envy. Even though I was genetically Chinese, I knew little about my ancestry and culture. Except that there was lucky money in red and gold envelopes, dragons were powerful and Chinese food at the Marco Polo in Vancouver with the now banned shark's fin soup was reserved for extra fancy occasions. In the 1960s, my mother nicknamed Chinese immigrants from Hong Kong "ding-a-lings", as they walked by her in public places. Likewise, Asians had nicknames for

Canadians of Chinese descent – bananas, Twinkies, sunny side up – all the same, yellow on the outside, but white on the inside.

Japan beckoned journalists and friends Christopher Donville and Laurie West, who both worked in Tokyo. Christopher worked for Bloomberg Business News. Laurie worked in a big newsroom for an Asian version of *Times* magazine. I vowed to take the opportunity to visit them, as my husband worked on contract from home to spend more time with Alisse and Cody. I took a Japanese night school course. I also watched early morning ACCESS Network conversational Japanese classes, which totally mesmerized Cody, who was just a toddler. In Calgary, Japanese restaurants flourished. Our family began a foodie love affair with Japanese food, starting with the grill of the Japanese Village Steak House, before upgrading to sushi and sashimi.

Then I entered the last stage of yellow fever. The cure, once and for all, was to travel to the Far East and see for myself. After months of planning and budgeting, I flew into Japan's Narita Airport, March 1993. I had been warned about culture shock, which struck me in an instant, along with its humid climate. I rode the bullet train from Narita Airport into downtown Tokyo. I felt like I landed on another planet at Tokyo's main train station – a labyrinth of underground passage ways and Kanji signage I could not comprehend, not one iota. Canada's population was established within Tokyo's perimeters. People swarmed around me like we were bees in a hive. Everybody was Japanese. It would be hours before I would see an Occidental person. It would be another two hours before I checked in to my base camp for the next week – a YMCA hotel in the Chiyoda-ku district of Tokyo.

The moment I checked into my room, it was obvious that Tokyo real estate was endangered. My room was about one hundred square feet. It efficiently housed a single bed, a narrow closet to hang up a few items and bookshelf desk with a chair that posted instructions as to what to do in case of an earthquake. I had to flex my body around the sink to sit on the toilet. The bathtub had only economy class leg room. At times like this, I'm glad I'm vertically challenged. I came to Japan to experience life the way locals lived. The hotel had a budget cafeteria, which served a bowl of steaming white rice at every meal, whether you asked for it or not. To the Japanese, steamed rice is sacred, like Mac'n'Cheese to a Canuck, and best consumed plain.

Informed of Tokyo's exorbitant food prices, that included ten dollars for a cup of coffee, I ate my breakfast and dinner at the YMCA hotel. I survived on peanut butter sandwiches and carrots for lunch. I ventured to try what looked like spring rolls from a hole in a wall establishment, only to find that my Western palate was no match for the Japanese. Ironically, I missed North American cuisine so badly. I actually ate a burger and fries from McDonald's, and again, winced at the taste of the seasoning of fries.

Frank was absolutely right about giving an hour to figure out which subway exit in Tokyo I needed, to take to find the building I needed to go to. A visit to the Akihabara electronics shopping district was a must. There I found a massive assortment of electronic goods, in the full spectrum of colours, sizes, shapes and features. The Akihabara rocked with sensory overload that included noise, crowds and the bustle of the vendors. Everything is negotiable, when it comes to price. I came home with a credit-card sized radio – a gadget that wasn't sold in America, but revered by the locals, who have an affinity for micro-sized items.

My trip to Japan was paid for by freelance writing assignments. I met Christopher and Laurie in Tokyo, as well as Toyota Corp. executives, and Canadian ex-patriates living and working in Japan. The Canadians, in particular, described how they acclimatized to the Japanese culture and business practices – paying attention to everything from bowing, handling business cards, and relationship building. I met a Canadian public relations consultant in Tokyo who earned over $200,000 a year in 1993, but after food, rent and other expenses, she was barely breaking even. Her tiny home office was large by Japanese standards. Her main living room, kitchen, washroom, and dining area measured around two hundred square feet. Her bedroom was in a loft, which she accessed by climbing up a ladder that folded up in the ceiling like an accordion when not in use.

The Canadian engineer who moved to Japan several years earlier was gainfully employed by Marubeni Corp. – a multinational Japanese conglomerate with a broad base of products and services. He was assigned to work in the electrical and industrial plant sector. He lived comfortably in Tokyo with his Japanese wife and young daughter. When dinner for four tallied several hundred dollars U.S., he shrugged his shoulders as this was to be expected. The four strawberries presented sushi style alone cost each person forty dollars. They were absolutely flawless in terms of size, shape, colour, firmness, and taste.

Logistics for this trip were tightly managed. I took a 30-minute train ride from downtown Tokyo to the Makuhari Messe, an area adjacent to the city of Chiba set up for international trade shows. My assignment was to report on FOODEX for Canadian grocery magazines. There were thousands of exhibits and samples to try, all for the Japanese palate, which would later make their way into Western fusion dishes, like black sesame ice cream or durian fruit.

Frank introduced me to the people at the Kikkoman Soy Sauce plant in Noda, in the Chiba prefecture. The visit was also a short train ride from Tokyo. The traditional Kikkoman soy sauce brewing process to manufacture soy sauce took a lot more patience than I thought it would – several months of fermentation of bacteria, yeast and proprietary cultures, followed by refining. Realizing this, I understand why the Japanese are so offended when Westerners pour soy sauce liberally over white rice.

In Osaka, I stayed at a Japanese businessman's hotel, for its economic pricing. The room was about twice the size of my room at the Tokyo YMCA hotel, but I had to share a communal washroom, as everybody shuffled around in bathrobes and slippers. A phone was provided in each room, with the number to call for a "massage". I dare not ask what the cost for that service was or if a happy ending was involved. The walls were dreadfully thin, as snoring from my neighbours was loudly heard.

While in Osaka, I met executives from Mitsubishi Corp., and toured their exhibition hall, which showcased technology under development – applying robotics and artificial intelligence. Unlike North American counterparts, Japanese manufacturers and technology-based companies invested heavily in research and development. Protocol was very heavy in Japanese culture, both at work and in family life. The Japanese executive who took me for dinner explained to me that he never had a direct conversation with his daughter. Instead, he would talk to his wife first, who in turn, would talk to their daughter. Whatever response she had would be communicated to him via his wife.

The young single Japanese women who work at the front desk of major international Japanese corporations are selected for their grace and beauty. They are the prettiest girls with the best skin, shiny black hair, and soft voices. Since the men worked such long hours, there was no time for them to find girlfriends. It was tradition that corporations did their best to find them potential mates. If the young lady at the front hadn't quit within a relatively short time (I heard a rumour that it was a year) because she got engaged, they fired her.

The corporation assumed that a match couldn't be made because she had a difficult personality.

I was not surprised by some of the cultural nuances I witnessed in Japan. Mariko, a 19-year-old Japanese student who we hosted in Calgary for six weeks the year before, told us that if a woman wasn't married off by 25, she was called a Christmas cake – inferring that she was past her expiry date. Unlike most of the Japanese students who went abroad to study English, Mariko didn't come from a wealthy family. She lived with her divorced mother. Remarriage was out of the question, Mariko claimed, as her mother felt slightly ostracized. Even so, Mariko said her independent mother was happier than most married women.

My go-to dinner in Japan was at the popular ramen and soba houses. Nothing was in English, but plastic replicas of the meals were displayed in the front windows. A plate of yakisoba and gyoza cost me around ten dollars. The green tea was added for next to nothing. It was easy to order. You just had to point at the plasticized meal you wanted. On one occasion, in a crowded soba house in Osaka, I had no other choice but to sit across the table from a middle-aged married Japanese woman. She spent her days shopping and dining alone at night. During the week, her husband lived in corporate managed housing in Tokyo. After a long day at the office, he was obliged to drink and eat with his co-workers. He came home on the weekends. On Saturdays, he went golfing with his co-workers. On Sundays, all he cared about was if his wife had finished his laundry. She accepted her life graciously, the way it was. I suspect, the highlight of her day was having dinner with me. She was so happy to practice her English with a foreigner. I was grateful to learn more about her life.

I ended my Japan travel with an overnight stay at Koyasan, Japan's holy mountain – a sanctuary of over one hundred Buddhist monasteries or ryokans, surrounded by eight other mountain peaks – all within a few hours from Osaka. You might say this was my 24-hour *"Eat, Pray, Love"* routine (minus the love part.) From the Osaka train station, I took the Midosuji subway line to the Namba subway station and transferred to the Nankai Railway Station. There I took a very delightful 90-minute train ride on the Rapid Express train to Goruraskashi train station, which weaved around curves of luscious green cedar-filled valleys. Then I took a breathtaking five-minute cable car up to Koyasan, which snugly hugged the side of the mountain. Finally, there was a 20-minute bus ride into the village, where I was dropped off at the first stop. I can't recall the name of the *ryokan*, but it was the one that was run by a Japanese woman who spoke flawless English.

When I arrived, I was greeted by a burly bald monk with a protruding bump just above his left eyebrow, tending to the front rock garden. He wore a black robe, which was secured around his waist and traditional Japanese thong sandals. It was like I arrived on set of a James Bond movie. I was fixated by the monk's bump above his left eyebrow. I'm sure he noticed me staring at his face. It was probably none of my business, but I hoped that the bump was not a cancer. The monk was at ease. He smiled at me and nodded. He was anticipating my arrival.

The monk directed me to visit the ryokan's owner, a sparingly slight woman in her sixties (or she could have been much older, as this was hard to tell from the serenity she exuded). She was dressed in a traditional silk Japanese kimono with her hair in an upswept hairdo. Her slightly powdered skin was flawless, her dark hair jet black and bright coloured lipstick popped from her face. As the travel guide confirmed, her English was perfect.

I sat across from her a table in a tatami floored room. It wasn't as uncomfortable as I imagined. There was a heater beneath the table. Our legs dangled into the space below the table. I had arrived mid-afternoon. I was in time to receive a frothing cup of green tea that her delicate hands deftly whisked up. The tea was served alongside memorable Koyasan red bean cakes. When I bit into them, my mouth savoured the sweetness and granular texture of the red beans. I made myself a mental note to go into town the next day to buy a box for myself. The sweetness of the red bean cakes complemented the bitterness of the matcha green tea.

I was the only guest at the ryokan that night, as the cherry blossoms wouldn't be arriving for another month. The tatami room that I slept in could have easily housed four more persons. I slept well under a mound of blankets. I was awoken at the crack of dawn by the sound of banging gongs and proceeded to spend the next hour in prayer and meditation. At the front of the temple were the monks, burning candles, and incense. What surprised me was that the monks chanted in Sanskrit, not Japanese. I just learned the origin of Japanese or Shinto Buddhists originated from the Himalayas, and travelled over to Japan via mainland China and Korea.

After the morning service, I was instructed a vegan breakfast would be promptly served traditional style in my room. I waited patiently, sitting cross-legged. The monk arrived and deftly arranged the pretty lacquered bowls and plates in order, to present the biggest visual impact. The food looked too good to eat. I had no idea there were so many ways to serve tofu with such an array of textures and sights. Some of the plates served a one morsel bite. I was not disappointed and felt very full. Dinner was even better and more elaborate. The edible sculpted coloured flowers took my breath away. If there is one meal to put on your life's bucket list, it would be dinner at a monastery on Koyasan.

I came back from Japan, happy to be home to see my family and the open landscapes. Alisse and Cody hardly noticed my absence. They hadn't changed, but I had. I was much richer inside, more than

I could have imagined. I was very fulfilled and thought that I had cured my yellow fever. But I was wrong.

The last and final stage of yellow fever is reinfection. This was spurred in part, by the postal code hopping of several friends. Christopher Donville moved from Tokyo to Singapore with his wife and new baby. Friend Anne Gover moved to Kuala Lumpar (Malayasia) for two years, as her engineer husband found a job there. John Patterson, an environmental advisor who I worked for at Western Oilfield Environmental Services a few years earlier, was living in Jakarta.

I couldn't say no. Especially, since I investigated and discovered that Singapore Airlines offered complimentary return airfare trips to Singapore for writers. I applied and was accepted. My strategy was to write my way around Singapore, Malaysia, and Indonesia. In 1994, I left Calgary again. To my delight, Singapore Airlines upgraded part of my flight to Raffles class, which is even better than first class. Raffles class is on the upper deck of the Jumbo jet that is accessible by a private staircase and reserved for only 24 persons. Champagne is immediately offered upon entry. Dinner was served on china with silverware and pressed linens, and always with a purple orchid. Every time someone used the washroom, a flight attendant ran in, wiped down the sink, and made sure the orchid hadn't been flushed down the sink. She even folded the tip of the toilet paper roll to a point.

The flight attendants at Singapore Airlines are known to be some of the prettiest in the world. They are hired for their beauty and grace. Corporate standards require a narrow age range, certain body measurements, including height and weight. I learned at Singapore Airlines' headquarters that each flight attendant is instructed on what makeup to wear to best flatter her facial features. Before each flight, it is mandatory for her to be weighed in. If she has tipped the scales a few pounds over her desired weight, she is grounded until she sheds those pounds.

All of my three friends in Southeast Asia had very different lifestyles, even though they were all married with children. Christopher and his family lived in a small two-bedroom apartment high rise with no air conditioning, but at least, they had a few floor fans. He only could jog at midnight, when it was cool enough. The jogging was essential to combat the battle of the flab. Singapore's gastronomical paradise is renowned for chili pepper crab served on the beach to pan dung cakes and hawker stall cuisine that embraced flavours of Southeast Asia. A westernized Asian city where people spoke English and had flush toilets, I found Singapore easy to adapt to. Anne, her husband and two children, tried living as normal a life as possible, renting a house in a residential neighbourhood in Kuala Lumpar, with a padlocked fence and front door. Then John and his family, lived as first-class citizens in a Third World country. John advised on environmental matters to the local government, hobnobbed with dignitaries and very important people in a league of their own. He lived in Bandung, an upscale town populated by ex-pats, west of Jakarta, in an estate home, with live-in servants. For one afternoon, I had a taste of this life. A businessman showed up at the Kartika Plaza Hotel with his chauffeur driven car to take me to the Canada Day celebration – a barbeque being held in the garden of the Canadian Consulate. In July 1994, there were one thousand Canadians in Indonesia, for work or travel. I was one of them.

When I arrived in Indonesia, I discovered a cultural mosaic – such diversity in language and culture amongst the six thousand plus inhabited islands there. The least sanitized of the three Southeast Asian countries I visited, Indonesia was definitely the most exotic. With my rudimentary Bahasa Indonesian language, I was accepted to my surprise as a local. Expecting to fork up USD$100 to $150 for the cab ride from the airport to the Kartika Plaza Hotel where I stayed, I was only charged around USD$25. On one ride between the Pasaraya Shopping Mall and the Kartika Plaza Hotel, the cab driver gave me spare change back.

I did the unthinkable in Jakarta, going to four meetings in a day. I was just grateful that meetings never lasted longer than an hour. Traveller's diarrhea kept me up all night for four nights and running to a bathroom every hour. While Imodium can stop the flow; it was these little yellow Chinese herbal pills that a Malaysian businessman prescribed would cure the cramps. Traffic was like my digestion – *amok*. I knew I was in a questionable neighbourhood, when the cab driver turned around to lock all the doors in the car.

I went to visit Miss Risa, an assistant for an international petroleum consulting company, which a Canadian geologist friend introduced me to. She was a happy young Indonesian gal, who opted for Western wear in a short dress and sleeveless top. I asked her what she did on Friday nights. She answered, "Drink beer and go dancing."

I was surprised as I knew she was Muslim. She giggled at me, "Yes, I'm Muslim, but not serious Muslim."

She also told me that she had a boyfriend she liked very much, but they couldn't marry. I asked her, "Why not?"

She replied, "He's Catholic."

I was surprised to hear this. While Indonesia is predominantly Muslim, there is a small percentage of the population that is Christian. They say that love conquers all. (Almost a decade later, I heard that Miss Risa got her wish and married her long-time boyfriend.)

I shadowed Miss Risa for a day. We went to a grocery store to buy several packages of dried noodles to give to family members of employees. With such economic disparity, some employers take on the responsibility of providing for extended non-working family members with food and other essentials. Everywhere I looked, I saw how close families were and how happy they seemed without the traipsing of the North American lifestyle. While North Americans will say how they think, I noticed that Indonesians often say how they feel.

I returned to Canada, enriched by all that I experienced and grateful to see my family. Back in Canada, I quickly wrote a plethora of articles on my travels. But as the year 2000 approached, I ran into a chronic case of writer's block. It became harder and harder for me to get excited about my assignments. After 16 years, I resigned from freelance writing. The timing coincided with the rise of the Internet, which was endangering the survival of writers.

PART FOUR –
THE COMEBACK

Chapter 39 –
Rednecks, Shotguns, and Shenanigans

I was fast approaching 50. Baby boomers who graduated in the 1970s, the chosen ones who floated to the top and were fortunate enough to invest prudently, were now primed for early retirement. As for the rest of us? The competition for work was serious. There were hordes of engineers looking for work, many who were better qualified than I and equipped with MBA degrees. Little did I know that what made me unique technically would be my saving grace.

Since the 1970s, Alberta became a breeding ground for environmental activism. This trend was here to stay. Legislation changed with the times. The *Canadian Environmental Protection Act* was established in 1999, so corporations would take things more seriously. Officers and directors were going to be held accountable for their misdemeanors – $1 million fines and/or sent to the slammer. Life behind bars didn't look so good. Conducting Phase I Site Assessments were in vogue. Consultants drove to sites to assess and report on environmental damage due to a plethora of oilfield operations, ranging from well sites to oilfield facilities or batteries. At the batteries, the oil produced from several wells is collected. Then the water and gas is removed from the oil before it gets pipelined or trucked to a refinery or processing plant.

Through my connections as a freelance writer, I heard about an up-and-coming small environmental services company. I procured a temporary part-time contract, at a deep discount rate compared to other engineers. I wasn't about to argue. The work led me into the heart of Alberta redneck country – first to Provost, where locals allegedly denied references to the Ku Klux Klan activity as published in *Maclean's* magazine and over to Smallville just a short drive west of Red Deer, where everyone's eyes were blue, except mine. Working on an environmental site cleanup near Provost, was a pretty good gig. Other than swatting flies at night in my room at the only motel in town. Listening to my co-worker's blasting of heavy metal music at the crack of dawn as his cure for hangovers, could be perceived as educational. I found the local contractor, a stocky redneck with tattoos up and down his arms and heavy gold chain around his neck, rather entertaining. "Nobody wants to talk to me in this town," as he listed off all the town wives he had bedded.

Working in Smallville on a contaminated site cleanup was a slightly different story. One of the operators, who insiders dubbed as "mentally unstable" called my boss and told him, "Whoever wrote that report better watch out."

I was informed not to worry. Such threats were just business as usual for colleagues, who stashed shotguns and baseball bats in the trunks of their vehicles. We knew, as we collected samples on the contaminated site that war was going to break out between the management and workers. I asked the operator, "Aren't you worried?"

He shrugged, "No, I got six guns."

I was flabbergasted, "Do you bring your guns to work?"

He replied, "Hell no, I use my knife. It's much faster." He raised his arm, as he demonstrated his throw.

Later, the technician who I supervised asked me why Billie Bob had six guns. I calmly replied, "One for every day of the week, except Sundays."

After Smallville, I was on the hunt for something less life threatening. The year was 2004. During 2003, the price of oil had risen above USD$30 per barrel, up from USD$25 per barrel and the trending indicated higher oil prices. Oilsands projects were in vogue. Fort McMurray was booming and so were engineering jobs.

Chapter 40 –
Corporate Whores, Hired Guns, and Free Agents

My Uncle Victor told me that the moment you go work for someone else, you stop living your life. You live someone else's dream. He was right. That's why as soon as he could, he started working for himself. In his late 20s, he owned and operated his first coffee shop – smack on 7ᵗʰ Avenue in downtown Calgary, across from the iconic Hudson's Bay building. He confessed to me that he knew nothing about the restaurant business, but he did everything. Make the coffee, serve the coffee and joke with the customers, so they'd come back for more. It was a tough go. Sometimes, he counted the pennies at the end of the day, to realize he may not have broken even. However, he had a spirit that could not be broken.

Decades later, his hard work paid off. Business excited him. He kept a stack of business guru books on his desk. All the wheeling and dealing turned a spark inside my Uncle Victor. When he talked about pie in the sky ventures, his face lit up like it was Christmas again. I wonder if that spark is genetic. As I wrote my last feature stories for *Alberta Construction* magazine, I felt an angst in my heart, as I watched the bulldozers and Caterpillars zig-zag cross my path. The stench of sweat, the dust and dirt didn't faze me. I missed the action. I missed seeing shit get done.

The powers that be must have read my mind. Not in my wildest dreams would I imagine that in 2004, I'd start my engineering comeback as the Health Safety and Environmental Lead Engineer for the Suncor Millennium Project. I worked for Bantrel, a global engineering, procurement and construction company. Unlike working for Fluor decades earlier, engineering projects were on steroids – the budgets were in the billions of dollars – a size that we once thought was ludicrous. The schedules were tight and clients pushed teams to the maximum. Personal computers hooked up to the company server. Offsite remote access had us chained to our jobs. There was no excuse not to work, even if you're sick or in recovery.

My uncle was right. I'd give up my lunch breaks, part of my evenings and weekends and sleep, to get the project done. On the other hand, I'd never seen myself so high up on the organization chart. It was like I just got the keys to a new Ferrari – money, power and as the guys would say, the big dick syndrome rolled into one. When I realized that I was the one who called the shots, I thought to myself, "Oh my God."

I was the third engineer in this position within three months. Two predecessors quit before me. My peers didn't think I would last six months. Some colleagues in jest were waging bets on my demise. I was one of the five lead technical engineers. I was assigned to ensure the engineering design met all health, safety and environmental technical specifications and regulations.

This time around as an engineer, human resources didn't ask me about birth control or sex. Women engineers were like dandelions, with Alberta leading the nation's highest concentration of women engineers per capita in the country at forty per cent. Long gone were the *Mad Men* martini lunches. Bantrel was a high paced work environment. There was no time for political correctness or apologies. If you couldn't hack the pressure, you could always leave. This work-place wasn't for whiners or spoon fed employees.

I was earning more than double what I earned working for an environmental services company. I supervised up to half a dozen junior engineers and technical support staff. Fort McMurray drew us masochists in like booze to alcoholics. Mega-projects had mega-challenges. We, engineers hell bent on problem solving, had a field day from day one. The major engineering, procurement and con-struction companies needed to have their in-house work flow pro-cesses and procedures in place. Everybody knew what was expected of them, what their deliverables are, and when they're due.

I would be lying to say that my husband and two children came first. My career came first now. Or at least that's what my ego led me to believe. I racked up man-hours during the week that pushed me into dire need of a 12-step program for workaholics. I got to dress up to go work again, except it wasn't necessary for me to look like a flight attendant. Dress pants and a black tailored shirt with low-heeled shoes were quite acceptable. Except for the most senior managers, very few men wore ties to work or even jackets.

My job demanded that I be efficient. I was. I rose early several times a week to swim lengths before breakfast with the family, then head off to the office. Sunday afternoon was spent in cooking up a storm to last at least until mid-week. Fridays were take-out Chinese food or pizza. Alisse and Cody were bribed to make their own beds, lunches, do their own laundry, and homework.

Fort McMurray was a magnet for workers, first from eastern Canada – Newfoundland, in particular with its chronic high unemployment rate – and later from just about any country around the globe – Columbia, Venezuela, Mexico, Iran, India, Nigeria, etc. The oil sands plants owned and operated by major multinational conglomerates ran 24 hours a day.

In 2004, Suncor alone churned out 200,000 barrels of oil daily, a significant increase in production from earlier years.

The Suncor Millennium Coker Unit project involved the engineering, design, procurement, and construction of additional Coker units to the Suncor site. The Coker is the process vessel in a refinery operation that separates the solid black heavy coke from the coveted lighter hydrocarbon liquids and gases; which would be further refined into desirable commercial grade products, like gasoline or jet fuel.

My first pilgrimage to Fort McMurray would be followed by many. The Suncor site was as large as the Calgary downtown core, several city blocks long and just as wide. I felt like I stepped into a science fiction movie. Heavy machinery pulsated, steam and exhaust released themselves into the atmosphere, accompanied by shrill and sometimes, hissing sounds. The Suncor site had washrooms, and a company cafeteria with a mini salad bar. Field washrooms kept their toilet seats down, double ply toilet paper was standard and feminine hygiene products were supplied by the company.

For a working mother, field work was a joy. I arrived to a clean hotel room after a 12-hour work day, all you can eat buffet dinner, cable television, and uninterrupted sleep. Over the next year, I would frequent Fort McMurray at least once a month. The ratio of men to women was ten to one. Single women claimed the available men in Fort McMurray were not ones they wanted to take home to Mom and Dad – a bit rough around the edges with missing front teeth from bar room brawls or missing digits from accidents. Many persons came for the money. Long-distance romance and separations made fidelity challenging. Fort McMurray had its dark side – crack and cocaine capital of the world. I came to work. I loved to learn. I was stoked.

It was a two-minute elevator ride up to the top deck of the Coker Unit – deemed one of the most hazardous places to be in the world, as bitumen boiled around 450 to 500°C. The concentration of volatile organic hydrocarbons in the atmosphere is so high that vehicles are banned in this area. Industrial grade tricycles are the mode of transport all year round. A ride up to the Coker Unit deck required procurement of safety orientation and numerous permits, mandatory sign-in and escort by a Suncor site representative. When the wind blew, the pin-jointed structure would sway accordingly. The permit centre, which resided by the Coker Unit, is in a massive concrete walled blast proof building, where doors slide open like a deep freezer.

In the winter, fog prevailed as valves frequently froze up, pump seals cracked and leaked making them subject to fires. Cell phones were banned from the site. Drug testing was mandatory as 80 per cent of incidents on site were drug- and alcohol-related and a pass on safety orientation was 100 per cent. In the spring, with torrential rains and snow melt, the ground turned into mud that sucked everybody's boots in. In the summer, tar sand beetles that grew up to several inches in length (but most of the buggers I encountered were only about an inch long) propelled themselves like drones to withering heights, smacked a windshield and spilled their guts or sank their teeth into some unsuspecting worker.

Suncor knew that a cohesive project team was necessary and planned accordingly. The first few weeks on the project were spent in team building. There would be no time when the floodgates opened, for identifying who's who in the zoo. Activities included building structures from spaghetti and marshmallows, Star Wars-themed skits without words, and of course, pizza and beer. One team member pranced around with two saucers to let us know that the aliens were coming. Three cost control engineers did a strip tease.

On the Bantrel team, six of us project lead engineers were born within a week in April. Just for the record – not that astrology means anything.

I credit my peers for my survival. I was the least experienced engineer on the technical team. Incorporating health, safety and environmental design features during engineering design was back then, a relatively new novel concept. The mechanical lead fondly referred to as "Uncle Cece" for both his technical and private advice, became my mentor, but more significantly; a motherlode of office politics. A veteran engineer from Saskatchewan, Uncle Cece had a stellar oil patch career both in Libya and in Alberta. Though mild-mannered and unassuming, he had a wicked sense of humour.

I wasn't the only female engineer on the team. Bantrel hired women engineers from around the globe, who could be found working in all engineering disciplines – mechanical, civil, electrical, structural, and process. There were also many engineering interns to supervise, including several female ones, who studied engineering because their fathers encouraged them. With all barriers broken for them, they graduated in 2006, landed plum jobs, married and lived and worked happily ever after.

A year later, the engineering design team got reassigned. The adrenalin levels fell. Depression reared its ugly side. My role on the Millennium Coker Unit was coming to an end. There didn't appear to be any work on other projects. Like everyone else, if we weren't laid off, we quit to find work elsewhere. In hindsight, working so closely together wasn't such a bad idea. Many of us became good friends and kept in touch years after the Bantrel project ended. All of us were gainfully employed until the next oil industry downturn crept upon us, starting in 2013. Today, some of us fantasize that the price of oil will shoot up to USD$200 per barrel and companies will be begging us to come work for a few grand a day, as free agents and not hired guns. We were once corporate whores, and vouched a pact, never again.

Chapter 41 –
Returning to the Dog

It would be exactly 20 years after my first layoff from Husky that I would return to work there in 2006. Things were definitely different. Affirmative action and women's equality prevailed, starting with the office vernacular. The four letter F-word was heard frequently from the mouths of the office babes, especially the one in procurement who occupied the window office across from me.

You didn't need to drive standard anymore. Field trucks now had automatic transmissions. Maybe, it was because of more stringent occupational health and safety regulations or aging prostate glands, the company made sure you didn't need to take a leak at the back of the truck. If there wasn't a toilet on site, there was a porta-potty available.

The photos of 1970s topless busty broads had fallen off the face of the planet. The floors of the field office were so pristine, that if you dare put a mark on them, some field operator was going to chew you out. Bringing donuts worked in the 1970s, but the field operators and construction crew were concerned about eating properly, shunning donuts for apples and other fresh fruit. Lloydminster citizens were not content being labelled as rednecks. They yearned for the traipsing of urban centres, with Starbucks Coffee, Winners, sushi and the latest New York fashions found in a hipster boutique. Yoga was not foreign to the field crew. Conversations on energy healing or *reiki* was commonplace, as aging field crew with chronic back pain were open to less invasive options than surgery.

Some married male engineers with small children at home shunned field work. Working men with stay-at-home wives were almost extinct. It was fashionable for women engineers of child-bearing age to make babies, with no negative repercussions. They kept their seniority and even worked part-time.

Most of what went on in the boardroom didn't concern me. I was hired as a contractor. All I cared about was that the company cheque didn't bounce, a field vehicle that didn't stall, close proximity to a Subway franchise for meals in Lloydminster and, one of those back suites at the Best Western with the big screen television.

I didn't even mind that they put me in an inside windowless office. I was back in the saddle, working for an oil company in project management. I was getting stuff done. That meant anything and everything. I worked on the engineering and construction of pipelines to connect wells to production facilities, which were primarily often comprised of separator packages, tanks and pumps, along with a whole bunch of measuring and monitoring equipment. Sometimes, I got to work on the installation of compressors at gas plants or whatever maintenance work was required. I was never bored. Each project was unique in its own challenges. Small was beautiful – fewer people to contend with, more flexibility with schedules and more control on costs. Besides working on projects in Lloydminster, I travelled to some of the most unforgettable places in Alberta and Saskatchewan, that don't have hotels likely to be found on TripAdvisor because there aren't any.

Only in one day cruising around Fox Creek, Alberta, did I see black silky bear cubs, a moose, and a beaver. Come winter, everything was covered in knee deep snow. We had a six-week opportunity to get four wells tied-in. The challenge was construction in muskeg. Below the frost line, earth is like split pea soup. Equipment is heavy and the earth shook when the trucks inched along. The dampness chills you to the core. Out of sight didn't necessarily mean out of mind. In fact, quite the opposite. Underneath our feet was a labyrinth of other pipelines that we needed to cross, including major gas pipelines that transported Canadian gas to south of the 49th parallel. The execution for the well tie-in and pipeline crossing took advanced months of planning and co-ordination. We also prayed for a cold winter, so that permafrost formed a layer dense and strong enough to hack the heavy equipment loads.

There were six of us peering down a six-foot diameter hole below the permafrost: the government representative, two other third-party representatives, Husky construction representatives, and myself. The boring contractor operator had started the pipeline bore meters away from the pipeline crossing. The hydro-vacuumed hole was quickly collapsing. There was a sigh of relief when we saw the Husky pipeline emerged and cross the other high pressurized pipeline, with adequate clearance. Another day we got to live. I was elated. So was the Newfie boring contractor. He raised his arms victoriously for a moment, then cursed profusely.

Chapter 42 –
Flangeheads, Pipers, and Speaking French

A woman engineer, of my vintage, quit engineering to become an accountant. That's when she had reached her boiling point. She laughed about the time when she worked on site for a premium grade asshole. He told one of the ditch diggers to run her off the site with the bulldozer. She laughed at them, because that hunk of junk moved at half the speed at which she walked. Sad but true, but I have nothing as graphic as a bulldozer chasing me to report.

Eddie, one of my bosses with good intentions dished out some advice before meeting Willy Wilcox. He was the big badass dude in charge of construction or more specifically, the construction superintendent who managed all the construction field representatives. My job as project manager was just to make sure everything got done from engineering and design to procurement and then to construction. Eddie suggested that I try to get along with Willy, notorious for bullying others and always getting his way. Eddie, a mild-mannered man of average height and stature, described Willy as some lake monster with multiple arms and multiple jaws, "He's a real big man and he talks with a real loud voice."

I didn't scare easily. I already worked in Grande Prairie with Big Brian, obviously a local prairie boy raised on free range antibiotic free meats and GMO-free grains. Big Brian stood about six foot seven inches tall and weighed almost three hundred pounds. Big Brian may have been large, but he certainly wasn't no monster. He was a soft-spoken man as docile as a pussycat. All the boys in Grande Prairie were like Big Brian, begging me to come visit them more often, to break their peaceful idyllic existence. I look at the photo Big Brian took of me in the field. I looked like some kid playing make believe in the fire-retardant coveralls, safety glasses, boots and hard hat. All the equipment I worked on was much bigger than the Lego I used to play with and, much more dangerous, too. I think Big Brian was disappointed that the corporation couldn't replace the ladders on the side of the oilfield tanks with something larger. All that oil equipment was designed to standards from the 1960s when the average male stood five foot eight inches and weighed 150 pounds.

When I got to meet Willy for the first time in Lloydminster, I figured I should just be myself. Willy gave me a body scan, as I trekked into his office dressed for the field with my Mark's Work Wearhouse construction boots and jeans. He sized me up like all of his potential victims and muttered, "Five foot four inches, 120 pounds."

I agreed. It looks like corporate work wear made me bigger than I was. I also eyeballed Willy, too. He wasn't the monster that Eddie made him out to be. A shining bald man with a weather-worn face and saggy jowls around his jaws, he wasn't even five foot seven inches tall. Relative to Big Brian, I thought he was kind of a puny guy.

Willy's biggest fear was that the project I was assigned was going to end up as another "cluster fuck." To be perfectly honest, it was already a cluster fuck. The details are irrelevant at this point. Let's just say, what had been done onsite needed to be corrected. That meant work for me. I could tell that it was hard for Willy not to curse in front of me, or shame me. In our first rendezvous, he only spoke French once. But after that, he was quite accustomed to me, and the f-bombs flew out of his mouth like the Fourth of July.

Small town people have nothing to talk about except themselves. Everybody knew about Willy. He was a welder who started off his oil patch career welding above ground piping together, fondly referred to as a "flangehead". He saw the opportunity in the corporate realm and made his way up to the highest position possible – the construction superintendent. Willy was cursed from the day he was born. His father was a son of a bitch. They say that Willy mellowed slightly in his later years, but his cursing could be heard from one end of the office to the next. When Willy's big black truck appeared on the horizon, the contractor's tall millennial field representative fled by my side and sought refuge in the field trailer, "You deal with him. I'm not."

Now I have no right to play psychologist, but I had a job to do. How do I teach an old dog like Willy, a new trick? I knew he already picked on the younger field guys who were better educated than him, equipped with technical school diploma or university degrees. He abused his construction site field representatives with threats of dismissal, just because he was having a bad day. Eddie made it very clear to me that the corporation better start constructing roads properly. It was my job to get this task done. No matter what.

Literally throwing tons of gravel on dirt every spring didn't cut it. By fall, the gravel roads had fallen apart and trucks were getting stuck. The geotechnical engineer I hired was a twitchy tall slender middle-aged man. Willy felt threatened by him from the moment their eyes met. He didn't like the idea that there was another person in his territory checking up on his men, procuring special measurements. Let alone someone who literally looked down on him. I convinced everybody we needed an intervention. I asked Yurgi, the geotechnical engineering manager, to prove to Willy why throwing gravel on dirt wasn't a road. I intentionally chose Yurgi. He was a barrel-chested man who stood as tall as Willy. With a ball busting voice and quick wit, Yurgi could put Willy in his place. I never told Willy that Yurgi had a Ph.D. in engineering. That would have thrown him off course.

It was a perfect day to educate Willy. The dirt was graded where a road was to be constructed. They threw gravel over the dirt. The depth of the gravel road was measured. A loaded dump truck ran over it a few times. The gravel road sunk, and it wasn't even raining. Everybody nodded in unison, except Willy. Yurgi walked over to Willy who had his arms crossed over his chest. They spoke eye to eye. Willy still didn't get it. Then it must have been divine intervention. Willy looked down in deep thought, when he saw the light and reluctantly agreed to greenlighting the construction program.

I came by a week later to see the finished roads. They were beautiful. Willy was ecstatic and issued an order that from that point on in time, all of the corporation's roads be constructed that way. The young civil construction representative earned respect. I got rewarded, too. I got assigned to another cluster fuck.

This time, it was working for a piper – those guys in the patch who are only concerned in pipeline construction. Joe Brown was a purebred cowboy from a village in central Alberta that I once drove into by mistake. Joe had a university degree on saving the planet and had a steady paycheque working for the government. He ended up in construction because in those good old days, a good representative hauled in $300K for working nine or ten months of the year. Joe's morning salutations began with, "Hey Nattalia, how the fuck are you?"

My politically correct response recognizing his culture would be, "Why Joe, I'm just fucking fine ... and what about you?"

It must have been those years of repression or maybe hazardous exposure levels to heavy metals and volatile organic compounds in the field that led me to very controlled outbursts of demonstrating my ability to speak "French". It was quite commonplace for those field boys to swear in the boardroom and a moment later, apologize profusely, because there was a lady present. However, even though I was their supervisor, I would not reprimand them but acknowledge their need to express themselves, "Yeah, I speak French, too."

They would smirk like little kids in kindergarten. Only once, did I lose it in front of the field crew.

I was livid because the construction department changed a field protocol referring that all pipeline crossing plans had to be signed off by the big boys in the Calgary. This decision was made on Wednesday, the day I left Calgary for Lloydminster. At the Thursday construction meeting, I was informed that the construction couldn't be kicked off until this document was in place. In front of those six feet plus 250-pound plus linebackers who comprised the construction contractors, I called up the Calgary man to get the form signed off. He answered me and then put me on hold. In the interim, I yelled out to him, knowing damned well he couldn't hear me, "Well, get off your fucking phone, and get back into your fucking office, and on to your fucking computer, and get me that fucking form ... so we can get the fucking construction started!"

All the construction contractors had grins on their faces that couldn't be wiped off, even if you ordered them to. Trust me on this. It was very hard, if not possible for me, to keep a straight face in front of them.

Chapter 43 –
Sociopaths, Psycho Bosses, and Narcissists

I decided not to become a doctor because I didn't want to be working with sick people. But I'm pretty sure I did. Like ice cream, I've licked them all, although not making any headway in taking them down – sociopaths, psycho bosses or narcissists. They came in all heights, shapes, sizes, colours, and genders. Predators some say. They have no boundaries with their victims. On first meet and greet, they love bomb you with words of praise, charismatic smile, and they tell you what you want to hear. You're so impressed with their finesse and charm, you're hooked into working for them. They promise you rich rewards, good times ahead, and unlimited career growth.

A good friend of mine, an ex-patriate engineer who worked in Calgary during the booms, decided he'd jump ship from working for an engineering consulting company to an oil company. I and others warned him that we didn't think that was a good idea, considering rumours flew that this particular oil company gave new meaning to hell. Their corporate culture was typical for major multinationals, bloated by procedures and bureaucracy. It's really only an act of some higher power and high oil prices that these businesses were profitable. My friend lasted six weeks. A fit and healthy man despite his girth, he told me that he was drinking a can of Coke in the

178

boardroom when his supervisor told him, "If I was you, that'd better be diet Coke."

The company valued the physical attractiveness of their employees, more so, than technical competence. Unfortunately, I don't think my friend fitted that mode. He eventually returned back home and worked for a small company that has accepted him for who he is.

At another place of employment of mine, there was a brilliant engineer who could do no wrong. He drew rave reviews from his superiors, giving stellar technical presentations at conferences and adding prestige to the corporation. But his bile was as vile as his temper. He changed his priorities on which projects should be completed first, as fast as millennial can snap a selfie. If he'd be given a spirit animal, Tyrannosaurus Rex comes to my mind.

While I worked at one employer, I ran into one of my university classmates – a civil engineering graduate who was content to hide in his office, cough up stress analysis on pipes and beams. He earned less coin than what I did in projects, but he conceded the stress of dealing with the psychos in the upper echelons wasn't worth it. But I, with my dysfunctional upbringing was the perfect person to receive the turds that flew my way from the monkeys who oversaw our work. In the post #MeToo era, some of them would have never gotten away with it. Senior management meetings were fit for the asylum. Some complained that there was too much testosterone and perhaps, one or two ladies present, could bring some civility to the boardroom. I beg to differ. When sociopaths, psycho bosses, and narcissists rise to the top, the only recourse for the victims is to quit.

Where do I start? I know. There was William the Great. He was a tall dark curly-haired man, who was going through an acrimonious divorce. (They all do, and of course, it's her fault and not theirs.) He squinted at you with his piercing eyes over his puffy bags and droopy eyelids. His coffee-stained teeth were covered in part by his overhanging moustache. It's the beginning of the project, the honeymoon phase when all team members are playing nice and no one is playing hard to get (yet). William is making us feel like we're gods and invincible. Our efforts will be rewarded with bonuses (not likely), lots of fun and parties. That latter never happened. William wore an artificially wide smile on his face. When in front of the client, he was on his best behaviour, as a total yes man.

Then like a bad dream come true, there's a major deviation on another project that William was responsible and professionally liable for. We're talking about a project that was completed and commissioned three years earlier. It didn't take much for all that hardware to go up in smoke in a matter of few days. The details as to the cause are irrelevant as that kept lawyers happy for years to come. The aftermath was tied up in the blame game. I didn't even work on the project. However, William felt entitled to make me the fall guy. We were reviewing some technical documents for the current project in front of a colleague, a cursing Brit. When William shredded my ego to a pulp in a 30-minute rant, the Brit made his own executive decision to find another job elsewhere. He was gone in 30 days.

I wasn't alone in being the recipient of William's tirades. An executive who avoided confrontation two levels higher than William, refused to attend boardroom meetings where William should have been drugged and put into a straightjacket right then and there. That never happened. I quit within 90 days of enduring his toxic outburst. William stayed behind and flourished. There was a rumour that a relative in a senior executive position covered for him. William was invincible. I wasn't. By the time I left, I was reduced to the self-esteem of a homeless person. I was lucky that I got out of that shit storm. Some of my male peers, in their early 50s would experience their first cardiac arrests. Then find out that the company cut them off with health care benefits and offered no form of compensation while they were in recovery.

Chapter 44 –
Networking, the Old Boys Club, and Power Lunches

Over lunch at a Calgary Women in Science and Technology Conference in the mid-1990s, I sat beside a woman engineering graduate who fretted about not getting her professional engineering status. I asked her, "Why not?"

She explained to me that they handed out the provincial engineering in training application forms during a fourth-year engineering class. She told me that she intended to fill in the form and mail it in; but first she had something more important to do. She had to get her hair done at the beauty parlour, as she had a hot date that night. She stuffed the form into her purse and completely forgot all about it.

Upon graduation, she married a fellow engineering classmate who became a very successful engineer. She stayed home until their children were old enough to attend primary school, and then returned to work in an engineering-related field. Decades later, she said, "How could I be so stupid? I already did the hard part on becoming an engineer. I got the engineering degree."

And her work experience would have qualified to become a licenced engineer. She said it. I didn't. There wasn't much for her to complain.

By the mere fact who she married, through her husband's contacts, she was one degree removed from the Old Boys Club.

When people don't succeed, they're quick to blame their lack of membership in the elusive and somewhat exclusive Old Boys Club. I'm not talking about the Petroleum Club, which wouldn't allow females to join until 1989. Or the Ranchmen's Club, located in a grandiose building across from the historic Peter Lougheed House, which also has women members, today. Neither of these places have the caché they used to. In the 2013 Calgary downturn, which some of us former working stiffs call "The Armageddon", as political leaders focus on carbon taxes and climate change, the oil industry just doesn't shine like it used to.

At a Christmas happy hour at The Petroleum Club in 2017, myself and a young female stockbroker observed that the majority of attendees in the lounge were under 40, diverse in colour, sizes, and shapes. The audacious young men wore ugly Christmas suits – red or green jackets and pants adorned with Santa faces or prancing reindeer. The diehard old boys who had seen better days wore tuxedos and their mature wives dressed in sequined gowns and Loubin spiked heels, discreetly entered the main entrance to attend their private soirees in the oak paneled back rooms.

In the early 1980s, my Uncle Victor took me and my visiting mother for dinner at the Petroleum Club. A dress code was in effect. I wore a silk shirt dress, but my mother to spite her brother, came plainly dressed, in neutral coloured sweater and slacks. Her well made up face, toned down deep rose lipstick with her sculpted eyebrows and signature green eyeshadow from her eyelids to eyebrows were her saving grace, as the maitre d' didn't send her home, like she secretly hoped. I'm sure the distinguished man in the black suit couldn't help notice the brilliant sparkle of the VS1 diamond stud earrings she wore – just like Catherine Deneuve in the N° 5 CHANEL poster.

Uncle Victor well aware that he was the only "Chinaman" in the dining room that evening – came only in his finest – a made to measure Hong Kong tailored dark navy silk suit, pressed white shirt with gold cufflinks and red silk power tie. Even though he never graduated from high school, he proved that being street smart and working ridiculously hard would make enough money to buy you status. He rattled off all the tony business clubs he had gone to. The Chicago Terminus City was his favourite.

Whether I attended an engineering technical society luncheon in a western themed room at the Petroleum Club or the chandelier loaded Crystal Ballroom at The Palliser Hotel, I became versed on what fork was used for what. Or after finishing the main course, how to place fork and knife across the plate to let the server know it could be picked up. And most important, always tip the coat check attendant, on the way out.

I never drank hard liquor, smoked cigars or played golf. I knew membership in the Old Boys Club was off limits. While visible minorities will shout it's their skin colour that limits their career options, I heard amongst Caucasians that ancestry was a factor, even though their hair was blonde and their eyes were blue. Friends of Polish and Ukrainian descent claimed they were discriminated against. They just didn't come from the right lineage.

In hindsight, as a freelance writer, I realized that I hinged on the fringe of the Old Boys Club – comprised of many Calgary presidents and CEO's. They knew who I was from my by-line in *Oilweek* and *The Globe and Mail*. I knew them personally because I got to chat with them from their penthouse office suites from the 40th floor or higher. Only when my tape recorder was turned off, that one CEO admitted he really didn't mean what he'd said during our interview. It is true that a lot of CEO's are born with silver spoons in their mouths. However, there is only so much that nepotism can do. I like to think that you'll get respect, when you earn it.

Only once in my engineering career did I go power lunching. It was just one of those things that was meant to occur. It was about 1:00 p.m. one September sunny day when I was standing at the corner of 8th Ave. SW and 7th St. SW ready to cross 8th Avenue when the traffic light turned red. Low and behold out of nowhere, I looked to my left and there stood, a fellow UBC engineering grad. We recognized each other, as he previously worked at Husky during the 1980s, too. In the two minutes waiting for the light to change, we exchanged small talk as to what was going on in our lives. He told me that he was working for this new startup. It was an oil sands company founded by Phil Svensson, another Husky engineer who also worked for Husky when we did. Coincidentally, Phil's new company was in the building adjacent to Husky's.

Weeks later, Phil and I met for lunch at the now defunct Quincy's – an established steak house with studded leather chairs, thick plush carpet and big brass glass front doors. In 1984, I recall meeting Phil for the first time at Husky's Lloydminster field office, wearing hard hat matted hair, a t-shirt, jeans and steel-toed boots. This time, Phil arrived from a high rollers meeting. He sported a tailored pin-striped suit, dress shirt with white cuffs and gold cufflinks. He was on top of his game, but aren't we all, when starting a new venture? Except this wasn't his first venture. He was cool, calm, and collected.

The Quincy's lunch was my first and only power lunch that I ever had with an oil company's CEO. I was more than ready. Phil ordered a club soda with cranberry juice. I ordered the same. He picked the steak salad, curbing down on the carbs. I ordered the same. He wore navy. I wore navy, too. His dress shirt was bankers pinstriped blue. Mine was just plain white. We chatted like two colleagues catching up. He knew about the death of our firstborn son Andre, but wanted to know if we had more children. I told him about our daughter and second son. He told me about his family, too – his wife who has stood by him through thick and thin and his grown-up children. We didn't talk about anything technical or work. It was more like two old women gossips. I left shaking hands, and he said to expect a call from one of his vice-presidents.

The following week, I was called in to meet his second right hand man. Shortly thereafter, I was hired. I don't believe neither of them ever read my resume.

Chapter 45 –
Queen Bees, Mean Girls, and Batshit Crazy Bitches

When the construction crew is having an altercation onsite, the two dudes will duel. They will curse and name call each other at the top of their lungs, flail their arms in a frenzy, while their steel-toed boots are firmly planted on the ground. From a third-party perspective, it looks like they may go after each other's jugular and I should keep 911 on my speed dial. After five minutes of this blood-raising nonsense, their voices turn hoarse, their arms stop flailing, and they return to work side by side, as if nothing occurred. Peace has been restored.

No matter what feminists have to say, men and women are different – physically, biologically and how they think and react to day-to-day situations. Take dating, for example. The guys have their bro-code. They (normally) don't steal away their best friend's girlfriend. Girls on the other hand, need to work on redemption in this matter. I was on a second date with a nice-looking high school math teacher. We were shooting pool with a group of friends – when mean girl teeters in on her stilettos, a skimpy midriff top and low-rise tight jeans. When she bent over on the pool table, her breasts fell out of her top and revealed a small tramp stamp above her jeans. Needless to say, I went home by myself. My date went home with mean girl.

In the traditional corporate world, handing the baton from one outgoing senior male executive to an up and coming one occurs through mentorship in a culture where boys have bonded over team sports. During the 1970s, when women began infiltrating corporate ranks, there were no women ahead of them to mentor them, so those women who made it were referred to as "Queen Bees". When I returned to Husky in the 2000s, my presence was not welcomed by Zena Warrior, a sharp-tongued woman engineer, who told my supervisor that she did not want me on the project. She was angry that the young male engineer transferred into another department without her blessing.

My supervisor, who felt caught in the middle, sent me the e-mail between him and Zena Warrior about me. Then he had the gall to e-mail Zena Warrior back, copying me on their correspondence, as if to crush the office ménage à trois. When I ran into Zena Warrior at the photocopier, I thought I would do what I learned best growing up as a visible minority. Act deaf and look stupid. Zena Warrior was quite flustered around me, apologizing profusely that she really didn't mean to say what she did in her e-mail. Her apology wasn't really necessary. Zena Warrior's diplomacy skills were somewhat underdeveloped. Besides, she was just a Queen Bee, not a certified batshit crazy bitch.

One employer I worked for took pride in having a women friendly workplace, had anti-harassment workplace policies in place and supervisors were to respect the clothing that women chose to wear to work. The women took it upon themselves to bring new meaning to casual Fridays. They pranced around the office with cleavage normally exposed at night clubs, skirts so high up their thighs if they bent over they'd expose their butt cheeks and flip flops – so low quality that they seemed fit for public showers.

With higher percentages of women in the workplace, office gossip can be toxic and catfights can become problematic. Some of us already know that women have a hard time of letting go. A guy will put his personal differences aside to get the job done. A gal will smile demurely at everyone around the boardroom table and when everybody's back is turned, pull out her claws and plot revenge. In a short time, I realized that it wasn't the company's freedom of dress policy that made it such a women friendly work place. It was the fact that workers were paid less than elsewhere. When I figured this out, I quit.

In the latter part of my engineering career, I witnessed women engineers of the post-Lepine massacre, leverage their sexuality to the nth degree, as repercussions for sexual harassment had serious consequences. A giggling woman engineer confessed that her boss couldn't say no to her. Whenever she needed budget approval, she unbuttoned the top two buttons of her blouse before she entered the boardroom.

A busty female junior engineer came to work without her over the shoulder boulder holders. She strutted around the office boldly flaunting her breasts underneath a flimsy low cut blouse – more appropriate for a Friday night dance floor, not a corporate office. She went to human resources and filed a sexual harassment claim against a senior supervisor. Then she came to me to commiserate her woes. I ordered her to dress like a professional. That tirade got her coming to work appropriately, but caused major opposition from a colleague who missed his, "Friday afternoon delight."

Chapter 46 –
Pretty People Privilege

Uncle Cece ranted to me about some hot chick in the office. But I would ask him what about me? Did he think I was hot? After much deliberation, he carefully selected his words as if he were examining a technical specification, "Well, Nattalia. You're attractive but not like so-and so."

What he meant to say that I wasn't unattractive. I am considered "cute". I sure was no pretty girl, like Megan Fox.

Pretty girls get asked out more often than not so pretty girls. Pretty girls make more money than not so pretty girls. Pretty girls marry pretty boys and end up having more pretty children. Pretty people have friends because everybody wants to hang out with pretty people. Not singular to engineering, pretty people privilege is viral. No wonder the beauty and cosmetics industry are a multi-billion industry.

I can't speak for myself as experiencing pretty people privilege, especially since I am vertically challenged. I will confide in having pretty people envy, on occasion. (Okay, maybe more than once.)

When Loretta King walked in for a job interview, everybody's heads turned. She was a five-foot-six redhead with stunning green eyes, a complexion that other women pay big bucks for and the womanly bod that God blessed her with. Her naturally wavy shoulder length hair framed her face like Picasso would embrace his subjects. She wore a two-piece off white snubbed linen suit with high heels. Her cherry rose lipstick matched her shiny freshly manicured cherry rose nails. It was unanimous that she would get a job offer before she opened her mouth. The senior managers were fighting over who got to take her for lunch. The sleazy wavy dark-haired macho one with the curly moustache got first dibs. A mature engineer, battling high blood pressure, voiced his opinion in a raspy voice, "You can't hire her. Guys are going to drop their pants!"

Years later, I met Loretta at a conference in the ladies' room. She declined the job offer from Fluor because she got multiple job offers to choose from. She never worked in the consulting world, opting to work for organizations higher up the food chain, like oil companies. Back then, it was the blue-eyed sheiks that ran Calgary's oil patch. She whined that her good looks got her undesirable attention from men, which she didn't appreciate. Harassment is the term she used. That was foreign to me, due to the lack of infrastructure on my banana shaped bod. Even with the inconvenience of men panting over her, she fared way better than most women engineers. She became a vice-president of an oil company by the time she was in her early 40s and retired way before the masses.

Then there was Candice. She was a 10, just like Bo Derek. She had a sultry voice that went with her sultry look. The head honcho turned up his nose to three other women engineers looking for work that summer – a brunette with brown eyes in a dark corporate suit, a blonde with smoldering hazel eyes, and a willowy mousy brown blonde. Then Candice strolled in, with the grace and ease we saw Kate Upton stroll along the beach in the movie *The Other Woman*. He found a 10. Who cares if he was supposedly a happily married man with two kids at home? He was going to hire her. He did. I had no idea which budget he found funds from. He was so desperate to see Candice again. I'm sure he would be willing to forgo part of his annual bonus just to hire her.

Her presence in the office tower caused quite the commotion. There was a steady stream of bachelors, going out of their way to say hi to her and ask if she needed any of their assistance. Candice is the only woman who I have met who I witnessed get picked up in the lunch buffet line. I've witnessed Candice crazy men holler at her from across a main Calgary thoroughfare, waving their arms feverishly to get her attention. She never dressed fancy or wore any makeup. Except for some frosty coloured lipstick. She even tied back her gorgeous hair into a pony tail and came to the office. However, that did nothing to tone down the heat she created every time she walked. In the washroom, Candice asked me if she should wear eyeglasses to work. I knew what look she was going for. I told her that I didn't think it would matter.

Some thirty years later, I ran into Candice at a conference. Her sultry voice was the same. She had chopped off her long hair and wore tortoise shell reading glasses. Even after having four children, her bod snapped back in shape like an elastic band, without even having to hit the gym. While I was slathering my face with expensive imported skincare products, Candice only used simple soap and water. Her face was wrinkle-free. She was easygoing as ever. The company that kept her gainfully employed over the years bent over backwards to keep her happy. They surprised her with a bed in her office when she was pregnant. Now with school-aged children, she negotiated time off to coincide with all school holidays, and summers off. She basically worked part-time from Monday to Friday, with no loss in seniority.

As we reminisced about the good old days, Candice confided in me that the head honcho who hired us in the mid-1980s asked her if she

191

was planning to get pregnant. That was ludicrous! She only worked on contract for six weeks. He also invited her for drinks after work, with no one else. (I had no idea.) At times like this, I am glad that I did not have a bod that could have been used as a weapon of mass destruction. I didn't have the heart to tell her that the guys asked me if Candice got good grades because of her looks. That doubt never crossed their minds about me.

Unlike my hotter female peers, I never experienced the #MeToo of the 1970s and 1980s stuff to the extent they did – as they were grabbed, poked, and stroked in inappropriate places at the office. One girlfriend said a visit to the president's office was no different than a Hollywood casting couch.

I attribute my lack of #MeToo woes to my diminutive stature, which made it more challenging to harass me. Or maybe, what others considered harassment, I only considered teasing. At Fluor, one messy haired structural engineer as affectionate and shapely as a Vietnamese pot-bellied pig suggested that he could slip into something more comfortable. I didn't push back because it was his idea. (As you know, it's very important to stroke the male ego.) I agreed. In his office cubicle, the universe had arranged his desk free of engineering drawings, which could become our passion pit. It was frisky Friday afternoon, around 3:00 p.m. Why not? There was one problem, the lighting. The overhead fluorescent lighting was not conducive to lovemaking. It was way too bright. So it was. My potential blushing suitor backed down. After all, we were on company time.

Chapter 47 –
In My Defense, Mercury is in Retrograde

In September 2012, I moved out of the matrimonial home and filed for divorce. By then, we had met with five different marriage counsellors. They say we end up marrying a person who is like one of our parents. Enough said. You already know my story.

Earlier that year, my Husky contract (hired to cover for a woman engineer on maternity leave) ended. On my last flight out of Lloydminster airport, security so familiar with my face dismissed looking at my photo ID as I boarded the plane. They smiled at me like an immediate family member expecting me to come back the following week for a family Sunday dinner. I shook my head, but they didn't believe me, "That's what you all say. You'll be back."

I also informed the car rental clerk that I wouldn't be reserving the gold Liberty Jeep anymore, too. He shrugged his shoulders. He assumed there would be another oilfield worker driving it.

I can't count the number of Calgary-Lloydminster flights I've made since 1984. Or hours spent at the airport waiting for flights delayed from Calgary for the fog to lift from Lloydminster. I don't recall all the vehicles I got to drive around in Lloydminster, including the mini-van with the almost flat tire, the luxury Cadillac SUV with so many features that even the car rental clerk couldn't figure out how to turn off the lights and the KIA with jammed locks. I wonder how many feet of roast chicken Subway sandwiches I've eaten, too.

There was some hesitation as I walked across the tarmac to board the Central Mountain Air plane to Calgary. I took a shoulder check, as I knew this would be my final flight. There were flashbacks as how I excited I was the first time I flew into Lloyd. I remember how far the airport was away from any signs of civilization. Today, Husky's operations could be clearly seen on the horizon.

Years earlier, I had found myself sitting across the table from Frank, an engaging man in his 60s, who pontificated well and maintained a sardonic sense of humour. I'd never seen an astrologer before, but another engineer friend who wishes to remain unnamed, swore that our lives were somewhat pre-determined at birth and mandated by the stars. Needless to say, this did sound hokey. While others may disagree, a reading from Frank was worth the price I paid. It wasn't the truth that I craved for. It was the feeling that I felt better leaving his crowded pawn shop furnished apartment.

A neighbour kindly referred me to Frank, swearing she would never see him again. (She did, many years later.) He didn't sugar coat his readings, which he cautioned us, were not to be taken seriously, but for entertainment only. With all the grief from work interacting with sociopathic narcissists (which I referred to as "assholes" in the 1970s), visiting Frank had therapeutic benefits and kept me off anti-crazy pills. The first time I saw Frank, I was on top of the game working as a contract engineer for two oil companies. I had no reason to complain.

He, on the other hand, made some pretty profound statements about Calgary's future. The oil industry was going down, "You know Detroit. The auto industry left town and never came back."

Of course, I didn't take Frank seriously. Come on. He listens to podcasts, votes for the political party contrary to mine and reads astrological charts.

After my gigs with the two oil companies ended, I found a job with a small engineering company hedging bets that it was going to land this lucrative contract with an international gas company. That never happened. In four months, we all got our walking papers and the company filed for bankruptcy. Then I landed my last and final job, on permanent staff for another engineering company focused on the oil and gas industry. Things looked much better at this company. Within a year, I was promoted to engineering manager. Three weeks later, I was terminated without cause. The year was 2014. Terminated without cause is the best way to terminate employees, so they can't claim unemployment insurance or sue for severance. I got it. This was my 12th and final involuntary termination out of 14 engineering jobs. What can I say? Frank was right. Mercury is in retrograde.

PART FIVE -
COWTOWN TO LA LA LAND

Chapter 48 –
Destiny

All the girls in the office were seeing Wanda, a Calgary psychic who worked out of an inner city office. Wanda didn't call herself a psychic, but a communications consultant. She is not entirely misleading. Think about it this way. She's got the password to advanced communications with those on the other side. Brittany went. Tamara went to Wanda religiously every three months. Simone went. In the name of female bonding and *Ya Ya Sisterhood*, it was my turn. I pushed the intercom to Wanda's office in a mid-century low rise building. The worn tiles in the foyer were broken. Paint chips had flaked off the door frame revealing raw metal. A hoarse voice answered, "Come on in."

A short woman who looked like she rolled off the wrong side of the bed greeted me. Trailing behind her was her mini-me, a small squat short-haired pooch that waddled in sync, yapping at Wanda, trying to get her attention, as the leash is wrapped accidentally around its right front leg. Wanda apologizes as she bends over to free her pooch, and ushers me into her office. She grabs a cigarette and exits, "I'll be back. I've got to take the dog for a pee, and I need a smoke."

It's a sunny day outside, but she has the metal venetian blinds snapped shut. The overhead fluorescent lights have been turned off.

There's an Oriental theme going on in her mystical garden palace. A tacky gold smiling Buddha sits at the base of a vintage ceramic lamp with a red black-fringed lampshade. A stick of incense burns. I watch its smoke snake its way upward through the musty air. Dog toys and a bed take up a corner of the office space.

I walk gingerly to the armless wooden chair with a thin chair pad. I sit down and dare not to press my arms too hard on the scratched card table that Wanda has thrown a large scarf over. A pile of dog-eared Tarot cards is neatly stacked to one side. A half-melted candle burns to the other side. I wait anxiously, like a virgin who sits in the doctor's examining room, waiting for her first pelvic exam with apprehension.

Wanda barges into her office. Her pooch is somewhat relieved. Surprisingly, the dog buckles down to business, saunters over to its bed, lays down and falls immediately to sleep. She flutters around her office, looking for her shawl, not realizing that it's on the chair opposite to me. She coughs and clears her throat. The smell of cigarette smoke hovers over her. She pulls the shawl over her slouched shoulders, complaining there's a draft in the room. Sitting still is hard for her. She's a Sagittarius, she tells me. If our life can't be read in the cards, it sure can be read from our horoscopes and birthdates.

"What do I do?" I ask Wanda.

She checks her cards face up, to make sure none have left the earth plane or something. Then she flips them over and hands them back to me. I am told to shuffle them and then pick a set number of cards. She reassures me that I will pick the right cards. There is a strange sound in the room. Wanda is elated.

"Hear that?" says Wanda.

"Hear what?" I ask.

"That sound!" she exclaims back to me.

"Oh, it sounds like they just turned on the ventilation system," I reply.

Wanda shakes her head defiantly, "No, that sound. That sound is spirit. They are here. Right now."

Her eyes dilate, and she looks straight into my eyes.

"How can you tell?" I ask.

She's a bit delirious, I admit. For a few seconds, there is tingling in my fingers I swear. I place the cards down on the table. Wanda is elated as she proceeds with my reading. The next forty minutes rolls by in a flash. Wanda is highly entertaining. She talks fast. The rattling in the ventilation ducts stops. I am sure the dust bunnies that were stuck got sucked out somewhere.

Wanda's dog wakes up and yawns, rises and stretches a downward facing dog yoga pose. My reading ends. I hand over the cash. Wanda counts the bills and stuffs them in her wallet. Her hands fumble for her cigarettes. There were only two things that I got from her message. One was that my days with my husband were numbered and that my future would be in Hollywood, with the stars. I wasn't surprised about my husband, but me working in Hollywood, sounds like a hallucination of some sort.

I pointed to myself and asked her, "I am an engineer. How do I end up in Hollywood?"

With one cigarette dangling from the corner of her mouth, she shook her head, "Listen, honey, I only tell you what I see. That's your problem."

I didn't really think much of Wanda's reading. I returned to the office and the girls were dying to know what I thought of Wanda's fortune telling skills. I shrugged my shoulders, "I guess she's okay. We shall see."

Career-wise, I was on top of my game. I was juggling two contracts, working for two oil companies which just happened to be connected side by side through a covered plus-15 skywalk. I was approaching another mid-life crisis, when I questioned how I had put aside my inner Hippie values on saving the planet and the principles of sustainability that I studied decades earlier. In 2003, the Canada Green Building Council was established. Years before the Alberta Chapter was established, I volunteered on the organizing committee. A local entrepreneur told me that it would look good on my resume. Sustainability was standard for the building industry, but still rather foreign to the oil and gas business.

You're right. I hadn't learned from previous mistakes. I still had not dealt with my co-dependence. I decided it was time to bring sustainability to the oil industry. I founded Sustainable Industrial Development for the 21st Century in 2009. Surfing the web drew me to Seattle, to learn more about sustainability. That summer, I travelled to Seattle and met their Director of Sustainability, a most cordial man, who arranged for sustainability tours on Seattle.

It was on a cement plant tour sponsored by a local chamber of commerce at the now closed down LaFarge Plant, that I met Bill Cody, a Los Angeles documentary producer and writer. Over the cubed cheddar cheese, pretzels and baby gherkin pickles, my loose lips got me in a kerfuffle with some candid remarks about Hollywood. Bill challenged me to come up with a pitch and write a screenplay. That began my journey into screenplay writing.

I wrote my first screenplay, *Bitumen Blues* – a woman tries to shut down an oil sands plant against the wishes of the president and an ex-former lover who wants to keep it running to close a takeover. I hired Mr. Cody to advise on the screenplay structure and for his comments. The next step was to enter screenplay contests. In 2014, *Bitumen Blues* won the Grand Prize in the Beverly Hills Screenplay Contest and attracted the attention of another Los Angeles-based producer, who took an interest in meeting aspiring screenplay writers. A year later, *Bitumen Blues* placed Official Selection in the New York Screenplay Contest and The Write Brothers Screenplay Competition of the Canada International Film Festival.

My second screenplay, *The Blonde Bitch on the Mountain*, was pure pleasure and inspired by the premise as what would a woman going through a divorce do, when she discovers that her ex-husband to be has posted a photo of his girlfriend on Facebook. I conducted an informal survey amongst random women how should it end for cheating husbands. Should he get killed? Or should he suffer? With absolute certainty, they replied, "Make him SUFFER."

I thought of leaving him homeless in Central Park would be a good answer. I wrote *The Blonde Bitch on the Mountain* for Hollywood – a timid woman goes on a wild goose chase to collect what is rightfully hers during her divorce. I wasn't the only one who thought *The Blonde Bitch on the Mountain* was funny. My second screenplay placed Official Selection in several Los Angeles based screenplay contests. With all its international locations, *The Blonde Bitch on the Mountain* was no match for independent film producers, typically, with budgets less than half a million dollars.

Then I found myself attending speed-dating pitching events in Los Angeles, under the belief that a newbie like me could actually find a well-heeled backer to produce my screenplays. I spent many a day standing in line with dreamers like me, hoping to hook up with that producer looking for screenplays to boot. At the former InkTip Networking Summit, the tables were arranged by genre – whether it be crime, drama, romance, thriller, horror, family, etc. Each table housed up to four producers. The event organizers always claimed that heavyweights from major Hollywood production studios would be there. The question was how do you sell four producers in less than five minutes, as the bells rang when your session ended? That was only enough time to find your table number, say your first and last name and where you came from, the name of your screenplay and the logline – the DNA of your screenplay is – who the good guy or girl is, the bad people are and the main conflict to overcome. Writing a logline sounds like a no-brainer, but in essence, it is the heart and soul of every great screenplay.

While I was in Los Angeles, a Calgary actor suggested I look up another actor friend of hers. I had a panic attack when I got a text from David, saying that his car broke down, and he was going home to fix it. He arrived on time, as he predicted. Following his father's footsteps, David studied mechanical engineering after high school. He also liked surfing, too. With his striking good looks, especially shirtless, he became a stunt performer. One thing led to another and he starred in some major roles during primetime television for decades. David would be the first of many actors and producers, who I would later meet in film and entertainment, with engineering degrees.

After attending several Hollywood pitching events, my friends from Los Angeles and I, became somewhat skeptical of the merits of Hollywood pitching events. Eventually, we realized that if we're going to get our scripts produced, we have to do it ourselves. In 2015, I put some money where my mouth was and produced *Spikes at Her Elbow* – a docudrama on what it was like working as a female engineer in Calgary in 1979. It was shot over one weekend at the end of November and then took another six months to finish the editing and post-production. There were over 40 persons involved in the production of *Spikes at Her Elbow*, which is available on-line at https://vucavu.com/en/winnipeg-film-group/2016/spikes-at-her-elbow.

My fear of not finding a Chinese-Canadian actor in her early 20s was unfounded. There were at least a dozen young women to choose from. Unlike my generation of women from the 1970s, these women were empowered, proud and assertive. Erica Ho, lead actor did a brilliant job, but as a millennial, found some of the lines hard to swallow … and some of the fashion choices from that era despicable.

At two private screenings, I observed that some people were angry (I really think they took the short film way too seriously), some people laughed and one of the lead actor's grandmother was in absolute tears. A programmer for a film festival was incredibly upset by the film as it reminded her of the Marc Lepine massacre and saw it unfit for public viewings. Broadcasters claim that they only want to show audiences relevant topics that matter, and are only interested in contemporary topic matters. Feminists lament that women are the victims; conditions today are not acceptable. If only they were around in the 1970s. I say women today have a lot to celebrate and be grateful for the women who did the grunt work back then for future generations of women.

I am amused that my journey to La La Land has been rather serendipitous, in the post-Harvey Weinstein era. Like the engineering profession when I started, I have no mentors and no relatives in the biz. Yet, despite my age (as actors like Anne Hathaway and Olivia Wilde, at least half my age, have been passed over for Hollywood roles for being "too old"), and where I've come from, I am proud and grateful for my accomplishments. I don't know where I'm going to end up. For now, I'm rather happy. Except for those Twitter, Facebook and Instagram posts on the red carpet wearing clothing from consignment stores with a delusional look of fame, working in the film and entertainment is not really that glamorous. Thumbs up, engineering was a lot more. To-date for me, film and entertainment has been long hours, lots of risk and uncertain return on investment. Nonetheless, I have enjoyed the process, the challenge and the life-long learning.

Chapter 49 –
Like Mother, Like Daughter

My relationship with my mother was a rocky one, even before I left Vancouver for Calgary in 1979. She was already slighted by me moving out during third year university. I would make annual and bi-annual pilgrimages to see my mother. She would argue that I was wasting my time and money to see her. We'd meet at her favourite restaurant, The White Spot, so she could chow down on a burger and fries. Her hyper energy burnt off any calories, as she remained slim forever. Within 30 minutes of conversation, she could not contain being civil with me, then lash out, as how bad her life was, what a victim she'd become and how much better her life would have been by not having us children or being married to our father. Out of desperation and abandonment, I held on tightly to our relationship. It was pathetic, but she was my mother.

My mother was only 41 when my father died, but she declined dating publicly until my grandmother died. Then, I hoped my mother would find some happiness in her life.

The man that my mother brought home was Charlie, a short smooth-talking man with a thin moustache who worked in the entertainment industry as a music promoter, arranging gigs for up and coming musicians. He would be the closest thing to Hollywood that my mother will ever get. He had the audacity of claiming he had something to do with discovering Karen Carpenter, but my Uncle Victor was far from impressed and thought he was full of crap. Charlie was the kind of guy that she never wanted her daughters to date. He had been married four times and his last wife, he claimed, wouldn't sign the divorce papers. He drank hard liquor, smoked cigarettes and lived in squalor somewhere. For the next ten years or so, my mother was happy. She had a boyfriend. They went bowling every week. He took her out for dinner. He taught her how to cook better. They went to the Chinese opera together. He left his toothbrush in her bathroom. He had a change or two of clothes in her closet. I didn't judge her, although one of my Aunties couldn't help snicker about my mother's boyfriend.

I was just so happy that my mother was enjoying herself. Until one day, she called me up. I asked her about Charlie. She replied curtly, "I dumped him."

Apparently, Charlie's fourth wife still hadn't signed the divorce papers. Or maybe Charlie never intended to marry my mother, after all.

My mother gave back to the community, something she learned from my grandmother. The Chinatown community was a tight one. My grandmother cooked extra food and gave it to the men who came from China and lived in rented rooms, working and sending what money they had left back home to China. When another woman in Chinatown was widowed with small children, my grandmother took it upon herself to help feed and clothe this woman, too.

My mother was no different. She volunteered to work in the hospitals as a translator for non-English speaking Chinese immigrants. In her later years, she was active in the retirement centre she lived in, organizing monthly birthday celebrations, modelling clothes for the geriatric set and making all the table decorations for birthdays. She also studied Qi Gong and taught it several times a week. In 2007, she received the BC Government Lighthouse Award for being an inspiration to the community. That was her Oscar red carpet moment.

It was Christmas Day in 2012 that I went to Vancouver to visit my mother. We shared Christmas dinner together at the retirement centre where she lived. This time she did not get angry with me for coming to Vancouver to visit her. She was making amends. All her internal struggles had melted away. Forgiveness had something to do with it. We hugged. She told me she loved me and after a lengthy visit, she invited me to come back and visit more often. That never happened. Dementia set in shortly after. She died a year later, fighting all the way to leave. She was my one and only female role model.

Chapter 50 –
I Ended Up Like My Mother

I vowed not to end up like my mother. I really tried hard not to. I grew my hair long. I never drank black coffee every morning, like she did. I never wore Revlon ruby red lipstick and green eye shadow from my eyelids to my eyebrows. I never wore three-inch spike heels with fuchsia pants like she did. I got a good education and wore a suit to work most days. I went to the gym several times a week and lifted weights, so I'd be really strong. I told myself positive affirmations daily. I really hoped that my marriage would have ended like those ones you read about in Hallmark greeting cards, where they lived happily ever after. Yes, dear, thank you for being such a loving husband, after forty years of pure bliss and joy. You are my best friend, confidant and lover. Of course, that never happened to me.

I practiced yoga and mindfulness meditation, while my mother became a Qi Gong master. I drive stick shift European cars. My mother drove automatic American brand cars. I only had three siblings. She had seven.

But I still did end up like my mother. She had white people food envy. So, did I. Only in her later years, she patronized The Ridge Garden Restaurant, which she claimed made the most authentic Cantonese Chinese food that her family made and without the monosodium glutamate. She'd call in a day in advance to make sure they'd prepare a tofu dish, ginger chicken, beef and green beans with black bean sauce and steamed cod. Recently, I confide in craving for a good fix of authentic Cantonese food, being eating around a round table with a Lazy Susan, surrounded by another eight or nine persons.

My mother was a patron of the arts, savouring Chinese opera and getting gussied up to go out for the evening. She didn't care if there was nobody to go with her. She went herself. I am pretty sure that my mother would have made a great movie star. She kept so many secrets, just like her mother did. She never lived long enough to see her daughter walk the red carpet. I know she would have been very happy to be in Hollywood.

In August 2017, I headed to Los Angeles to attend a red carpet to receive an award for a family feature screenplay. People from all over the world flew into La La Land to celebrate our achievements. As I sat down amongst the masses in the theatre, I clapped and cheered the winners on. It was a time of our lives, when everybody is glowing, no matter how jetlagged we were. It was a victorious time. We all had to overcome our doubts, anxieties and challenges to write that perfect screenplay or produce that amazing film that rips you to the core and makes your heart feel stuff that it's never felt before. There were tears of joy and emotion. I was having a good time just watching others receive their awards.

Then halfway through the ceremony, they called my name. "What? I won? Really?"

I felt surprise. I got up from my seat and teetered on my heels to receive my award. Everybody clapped. I felt everyone's love. It was such an honour. I smiled, like the way Emma Stone would. Weeks earlier, I showed the Sephora clerk an Emma Stone photo and asked if I could wear that same lipstick colour Emma was wearing.

Nothing mattered. Everything that I learned from being an engineer was dissipated into thin air. I blew forty dollars on a tube of YSL lipstick that I would only use once.

A week later, I came home and looked at the photo and realized OMG. I look like my mother. I had that look. It was her look in my eyes. I looked like my mind was in the ozone layer or maybe I was just having an out of body experience. For once in my life, I also realized that I wasn't pretty. I wasn't ugly. I AM BEAUTIFUL.

Epilogue

Lady with the Iron Ring was written to recognize women engineering students and graduates of the 1970s – who found themselves going against societal convention, the men who encouraged them and the men who were silenced for going beyond the call of duty to ease their transition into the once white-male dominated profession of engineering.

I chose to study engineering in 1974. As a pioneer, I had no idea of the strength, perseverance, hard work and guts required. There was no clear-cut path, no pats on one's back or adulation. There were challenges, trick turns, roadblocks, and barriers to jump over. I could have quit. I could have quit so many times. But I didn't. The truth is I enjoy working as an engineer, and I also enjoy working in film and entertainment.

As for what happens next in my life? Who knows. Some friends would like to see my memoir be transformed into a film or television project. If you agree, please follow, like and share www.fb.me/LadywiththeIronRing.

Thank you for reading my memoir. Everyone we meet is a blessing or a lesson. The ones who torment us the most give us the greatest opportunities to grow. Experiencing the hurt, the pain, the anger, the resentment and the fear, only brings us closer to acceptance, gratitude, joy, forgiveness, and love. I have no regrets. I wish you well.

Acknowledgements

I thank Denise Chong, Writer-In-Residence at the University of Calgary 2017-2018 for her comments and encouragement when writing this book. I also thank writers and journalists, who have guided me with my writing, including but not limited to Mike Byfield, Christopher Donville, Sophie Kneisel, Rick Spence, and Dave Pyette.

I am grateful for the countless professional engineers who have made me laugh and encouraged me in this journey called life. They include but are not limited to Mike Brawn, Edward DeGraaf, Doug Krupp, Ed Lawrence, and Cecil Chabot.

Let us recognize and acknowledge the professional engineers who have made engineering more inclusive for women including UBC Professor Dr. W.K. Oldham; Patrick J. Quinn, and the late Claudette Mackay-Lassonde.

Above all, I am grateful for the 13 other women who enrolled in engineering at UBC in 1974; and the four women who graduated from engineering with me in 1978 – Barb Dabrowski, Katie Hunter, Barb Tratch and Nancy Van Allen. The presence of my female peers alone was enough encouragement for me to graduate.

I am so fortunate for the love of Alisse and Cody, my two children who have brought joy and happiness into my life. I will also be eternally grateful for family members who will speak to me, after reading this book, too. They will be the ones with the humour gene.

Selected Bibliography

Engineers Canada. "Women in Engineering"
https://engineerscanada.ca/diversity/women-in-engineering

Harragan, Betty Lehan. "Games Mom Never Taught You" Warner Books 1977

Hopper, Tristin. "B.C. property titles bear reminders of a time when race-based covenants kept neighbourhoods white" *National Post* May 16, 2014

Lea, Nattalia. "Oil Patch Engineers – the Female Factor" *Alberta Business* May 1988

McKay, Shona. "Boys' Club" *Globe and Mail Report on Business* magazine September 1992

Molloy, John T. "Dress for Success" Warner Books 1975

Library and Archives Canada. "Sixth Census of Canada, 1921" Ottawa, Ontario, Canada: Library and Archives Canada, 2013

Library and Archives Canada. "Immigrants from China, 1885 – 1949" Ottawa, Ontario, Canada: Library and Archives Canada, 2013

Rudder, Christian. "Surprising Statistics About Hot People Versus Ugly People" posted on *OkCupid's OkTrends* on Jan. 13, 2011

Simon Fraser University. "Vancouver Chinatown" https://sfu.ca/chinese-canadian-history/vancouver_chinatown.en.html

University of British Columbia. "Chinese Canadian Stories: Uncommon Histories from a Common Past." October 13, 2010 http://chinesecanadian.ubc.ca

Vancouver Public Library. "British Columbia City
C=Directories 1860-1955" https://www.vpl.ca/digital-libary/
british-columbia-city-directories

Wikipedia. "History of women in engineering"
https://en.wikipedia.org/wiki/History_of_women_in_engineering

Wikipedia. "Queen bee syndrome"
https://en.wikipedia.org/wiki/Queen_bee_syndrome

Wikipedia. "2000s energy crisis"
https://en.wikipedia.org/wiki/2000s_energy_crisis

Printed in Canada